AF735322

SMALL-SCALE MANUFACTURING INDUSTRY IN IRELAND

Copies of this paper may be obtained from The Economic and Social Research Institute (Limited Company No. 18269). Registered Office: 4 Burlington Road, Dublin 4.

Price IR£7.50

(Special rate for students IR£3.75)

Kieran A. Kennedy is a Research Professor and the Director of The Economic and Social Research Institute and Tom Healy is a former Research Assistant at ESRI. The paper is accepted by the Institute which is not responsible for either the content or the views expressed therein.

SMALL-SCALE MANUFACTURING INDUSTRY IN IRELAND

KIERAN A. KENNEDY and TOM HEALY
(assisted by J. Bergin, T. Callan and P. McNutt)

DUBLIN, 1985

ISBN 0 7070 0076 9

Acknowledgements

The authors would like to express their deep gratitude to the following individuals for their comments on earlier drafts: Professor D. Conniffe (The Economic and Social Research Institute), Professor D. McAleese (Trinity College, Dublin), Dr. P. O'Farrell (University of Wales Institute of Science and Technology), Dr. E. O'Malley (ESRI), Ms. D. McHugh (ESRI), Mr. K. McGowan (Industrial Development Authority), Dr. J. McKeon (IDA) and an external referee. Valuable comments were also received from a number of organisations, notably the Department of Finance, the Department of Industry, Trade, Commerce and Tourism, the Industrial Development Authority, the National Board for Science and Technology and Shannon Free Airport Development Company. The Industrial Development Authority gave generous access to data sources as well as a financial contribution to the cost of the research. The authors also wish to record their gratitude to the ESRI secretarial and editorial staff, and in particular Florence O'Sullivan, Bernadette Payne and Mary McElhone. None of these above mentioned individuals or organisations, however, bears responsibility for the content of the paper which rests solely with the authors.

CONTENTS

LIST OF TABLES

Table		*Page*
9.8	Comparison of Anticipated and Actual Grant Rate and Capital Intensity in SIP Cohorts	146
9.9	Comparison of Conversion Rates and Grant Cost per Job in SIP and Other IDA Programmes	147
9.10	Annual Closure Rates 1973-79 for SIP Establishments Approved in the Period 1967-78	149
9.11	Employment Losses due to SIP Closures 1973-79 as a Percentage of Total SIP Employment by Industry Group	151

APPENDIX TABLE

General Summary

Economic doctrine, like most aspects of human thought and behaviour, is prone to changes of fashion. This is very evident in regard to views on the size structure of industry most suited to economic progress. For long the dominant emphasis was on large-scale production units, which were thought to be more efficient and innovative. But in the last decade or so the pendulum has swung back, and in many countries small-scale industry is looked to increasingly to produce industrial dynamism and more jobs. All fashion changes, however, run the risk of swinging too far, with inevitable disappointment when over-optimistic expectations fail to materialise. To provide a solid basis for policy action, therefore, it is important to assess the weaknesses as well as the strengths of small industry. In that way we can hope to arrive at a realistic definition of the role of small industry and avoid groundless euphoria.

The present study attempts to provide such an assessment for small-scale manufacturing industry in Ireland. The study is divided into three parts. Part I reviews evidence for Ireland and for other countries on the importance of small industry, its performance in carrying out the different functions of business, and its capacity for growth. Consideration is also given to the significance of developing new enterprise, and to the policy stance adopted in other countries towards small industry. Part II examines the operation of the Small Industry Programme initiated by the Industrial Development Authority in 1967. Part III discusses the policy implications for Ireland and makes proposals about future strategy in regard to the development of small industry.

Characteristics of Small Industry

The structure of manufacturing in all Western economies is characterised by a large number of small firms. Defining small establishments as those with less than 50 persons employed, they account for over three-quarters of the total number of manufacturing establishments in Ireland and in most other countries. The share of small establishments in total manufacturing employment, however, is much smaller: in Ireland it amounts to just under 25 per cent. Their share in total output or total investment in manufacturing is less still, since small firms usually have a lower than average level of output per head and a relatively low degree of capital intensity. By reference to other countries at a similar stage of development, Ireland does not have a particularly large small firm sector. At the other end of the size distribution, Ireland also has a low share of manufacturing employment in large establishments, and the average size of these large establishments is small by international standards. More so than in most countries, therefore, manufacturing employment is concentrated in middle-sized establishments (50–500 employees).

In all countries, the generality of small firms are technologically unsophisticated. Nevertheless, in some countries, especially the United States, a minority of the small firms can and do make a significant contribution to technological change, particularly at the inventive stages of new products and processes. Innovative small firms tend to be concentrated at the early stages of the product cycle when technological development is rapid and fluid. The great bulk of small firms serve local markets and encounter considerable problems in exporting. Finance is almost always a significant problem for small firms, especially new small firms, partly because of their own poor financial control and expertise. But, in addition, financiers are often reluctant to invest in, or lend to, businesses which are perceived to have no track record, offer little security and are vulnerable to competition from large firms or other small firms. Such perceptions are not without foundation, of course, but if applied in a blanket manner, they can result in a "finance gap" even for viable small firms; and virtually all countries have found it necessary to take special measures to improve access to finance for small firms.

Typically the very small firm is managed solely by the owner, who may also spend some of his time working on the factory floor. This can lead to serious management inadequacies in regard to certain functions in which the owner/manager has no expertise. Moreover, the manager is often too busy to take time off for management courses. As the small firms grow, a more specialised managerial structure becomes essential, but many small firms fail to make the necessary adaptation — sometimes because of the reluctance of the owner to dilute control through delegation.

Why then, despite facing disadvantages in the performance of one or more of the key business functions, does a significant small firm sector continue? There are several reasons. In some activities economies of scale are not significant and it is possible to compete effectively although producing only on a small scale. Such activities tend to be those with low capital intensity, low fixed costs, batch production techniques, non-repetitive tasks, or tasks requiring personalised skills. Big firms may find it efficient to farm out such tasks to small sub-suppliers. Small firms are more likely to flourish where markets are localised or where there is a high degree of personalised service in delivery. Moreover, the industrial structure is constantly changing, and the small firm sector will comprise at any given time new firms on the way up and old firms in decline. The small firm sector is, therefore, one of great diversity and its composition is continually changing.

This high degree of variability extends to many aspects of the performance of small firms. Though there is no conclusive evidence that small firms are less profitable than large on the average, their profits fluctuate more. Since small firms are engaged in a more limited range of production, they are less able to offset losses on some lines against profits on others. Because of the pervasive nature of unemployment in recent years, much interest has centred on the employment performance of small firms. Again the evidence points to greater variability in employment changes in small firms than in large. But the evidence also shows that, on average, employment growth has been relatively greater in small firms. This evidence, however, needs to be interpreted with care. Even though small firms are a better bet for employment growth than large firms, nevertheless they account for only a fraction of total employment. If employment in larger firms is static or declining, it is therefore most unlikely that a satisfactory overall employment increase can be achieved solely from employment growth in small firms.

The Small Industry Programme

The Small Industry Programme (SIP), operated by the IDA (and by SFADCo in the mid-west region), is the major national programme directed towards the development of small manufacturing industry in Ireland. Part II of this study describes the evolution of the SIP, which was begun in 1967, and examines the operation of the scheme. The SIP approved establishments are overwhelmingly indigenous and are heavily concentrated in the metals and engineering and wood and furniture industrial groups. The growth of employment in SIP approved establishments has been significant, and is widely dispersed regionally. Some firms have shown a capacity to grow into larger firms, but they are as yet a very small minority. Though SIP firms receive a higher rate of capital grant as a percentage of investment, the grant cost per realised job was considerably less than for large firms covered under the New Industry Programme. This was because more of the planned jobs were converted into actuality, and because the capital investment per worker was lower.

In evaluating such findings, however, it should again be borne in mind that there is a measure of interdependence between the different components of industry. Without the development of larger firms, many of the new small firms would not be possible. Some of them supply inputs direct to the larger firms, while many others depend on the markets arising from the increased activity and incomes generated by larger industry. Only a minority of the small firms would be able to enter and survive on the basis of export markets. In that sense small industry is the more dependent component. But though large firms might be able to get their sub-supplies from abroad, nevertheless the existence of efficient sub-supply activities near to hand can also be a considerable help in improving the environment for developing larger industry. Hence in considering overall industrial strategy, the interrelationships between different programmes must be considered.

Policy Implications

The industrial development of Ireland is far from complete, and it would be quite wrong to think it can be completed exclusively by small firm development. On the contrary, the structure of Irish industry is such that the chief focus of industrial strategy must be the building of more large companies. There are major economies of scale in the production of many products; important innovations, as distinct from inventions, generally require large resources; the export marketing capability of small firms is limited unless they can combine; and large firms are often needed to provide markets for small.

But there is no reason why this approach should not continue to be accompanied by a complementary strategy for small firm development. Indeed there are good reasons for doing so in recognition of the potential contribution of small industry to employment creation, the development of new enterprise, regional policy, and meeting the sub-supply needs of larger enterprises. Moreover, the outlook and the instruments needed to nurture new or small enterprises require a distinctive approach from that appropriate to larger firms. As one writer (Bannock 1981) put it, "to treat small firms in the same way as large is usually, in fact, to discriminate against them."

The present study goes further and recommends the adoption of an explicit two-tiered approach to small industry, which would recognise the distinction within the small firm sector itself between the great majority of firms likely to continue small, and the minority capable of becoming large. The first tier, which would be operated regionally, would be concerned with raising efficiency in a wide range of small firms in the interests of employment creation, realising sub-supply opportunities and regional policy. The second tier, which would be operated nationally, would seek to identify and develop a selected number of new and existing small firms with rapid growth prospects and capabilities. The different policy approach applicable to the two tiers is sketched. Some safeguards are also suggested to minimise the dangers of a proliferation of agencies and schemes dealing with small firms.

Chapter 1

INTRODUCTION

Prior to the last decade or so, small industry received little attention in many industrialised countries. Among the major reasons for this neglect is the fact that small firms, although by far the most numerous in all Western countries, often account for a relatively small share of output and employment — and up to the 1970s this share was declining in many countries. It was widely assumed that large-sized firms were more efficient. Moreover, with the exception of agriculture, the small firm sector is poorly organised as a pressure group. This in turn is largely influenced by the diversity of small firms, their spirit of independence, and their limited management resources which are often too busy for pressure group activity. Also such pressure groups as do exist are biased to the extent that they cover only those already established rather than potential entrants.

In recent years, however, there is much more interest in small industry, not only in newly-industrialising countries, but in the developed countries as well. To some extent, the theme "small is beautiful" originated as a reaction to the growing concentration of power in centralised government and in large industrial conglomerates. Quite apart from the social concerns voiced about such developments, there were fears about possible adverse effects on the pace of innovations arising from the ever increasing bureaucracy in both the public and private sectors. Confirmation of these views has been claimed from the fact that in recent years some of the largest firms in traditional industries in the developed countries (e.g., shipbuilding, steel, textiles and automobiles) have been in decline, whereas many of the "high technology" activities have emerged from much smaller new enterprises. More generally, the intractable combination of unemployment and inflation that has persisted since the early 1970s, has convinced some observers that new industrial structures need to be developed, in which smaller-scale, more participative, industrial units would figure more prominently.

In underdeveloped countries, the relatively greater interest in small industry is substantially focused on the need to secure economic development. The encouragement of new local industrial enterprise, which is in scarce supply (Storey 1982) is seen as an important step on the way to development, and such enterprise must usually begin on a small scale. Ireland also, despite its comparatively high level of income *per capita*, exhibits significant features of underdevelopment, including a poor response from indigenous manufacturing enterprise. With increased international competition for a diminished flow of foreign enterprise, the need to develop such indigenous enterprise, together with pressing jobs requirements, have attracted greater attention to the potentialities of small industry in Ireland.

Unfortunately, despite the generally increasing interest in small industry, the subject remains, as the Bolton Report (1971, xv) puts it, "little researched and poorly documented". An undue proportion of the growing international literature on the topic tends to be uncritically promotional, abounding in phrases such as "small acorns into large oaks grow". In this study we try to make a realistic assessment of the strengths and weaknesses of small manufacturing industry, with a view to providing a more informed basis for policy measures.

In this introductory chapter, we first discuss the definitional and measurement problems that arise, and then outline in greater detail what will be covered in the rest of the study.

1 Definitional and Measurement Problems

The Unit of Observation

When we speak of small manufacturing industry, the question immediately arises as to what is the appropriate unit of observation. Although several possible units of observation could be used, the literature on the size distribution of industry generally concentrates on two — the *establishment* and the *firm*. The establishment is the unit of production of a homogeneous activity; while the firm is the unit of common ownership or control of one or more establishments, and is often taken as synonymous with the term *enterprise*.

These definitions may be clear in principle, but a number of difficulties arise in translating them into practice. Consider first the case of the establishment. What if two homogeneous activities are carried out at different places, or two disparate activities at the one place? The general practice of the Irish Central Statistics Office (CSO) in such cases is as follows:

> Where two or more factories owned by the same enterprise, are engaged in the same type of production activity each is considered to be a separate establishment if they have separate locations. Where two or more distinct industries are carried on at the same location by the same enterprise each distinct industry is considered to be a separate establishment (CSO, 1973, p. 2).

It is clear from this that the number of establishments that will be identified depends not only on the number of locations at which production is carried out — a straightforward matter — but also on the classification of industry — a rather more arbitrary matter. Thus, for example, in the Census of Industrial Production (CIP) industry "manufacture of paper and paper products", a factory producing paper, cardboard boxes, packing products, stationery and wallcoverings would be classed as one establishment. If a finer industrial classification were used, however, each of these activities would constitute a separate industry, and adherence to the requirement outlined above would involve the identification of five seperate establishments in the one factory. In fact, the position is even more arbitrary than this since the application of the requirement is, as the CSO put it, "in practice determined by the existence of separate records or the possibility of making separate estimates" (CSO 1973, p. 2). It is important to bear these points in mind in any analysis of the data. This is especially so in international comparisons, since even if the different countries all used the identical concept of establishment, lack of comparability may still arise because of the way the concept is applied in practice.

Difficulties also arise with the concepts of the firm or enterprise. A Census of Industrial Enterprises has been undertaken by the CSO in each year since 1975 in response to EEC directives. An enterprise is defined as the smallest legally autonomous unit, so that in a group comprised of a number of companies, each individual company is generally treated as a separate enterprise. Thus if the term "firm" is used to denote the totality of activity in common ownership or control, it is not synonymous with this concept of "enterprise", since several enterprises could be part of a single firm. Comparison on the basis of such data, therefore, can be affected by changes in legal practice over time or across countries. In addition, there are difficulties in relating either enterprise or firm data to particular industrial sectors. In the CSO enterprise census, only enterprises wholly or mainly engaged in industrial activity are included, so that industrial activity undertaken by enterprises not classified as industrial is excluded. Conversely, many industrial enterprises are engaged to some degree in non-industrial activity, which *is* included.

Which is the more appropriate unit of observation in a study of small manufacturing industry? The answer depends on the nature of the particular enquiry. Take, for example, economies of scale. If we are considering economies of scale in production alone, the establishment would generally be the most appropriate unit. When considering economies of scale in purchasing, marketing, finance and technology, however, the most relevant scale unit may be related not to the size of the establishment or even to the size of the firm in a given industry, but to the overall scale of the firm in all activities — a factor that greatly complicates such analyses, since the very large firms may be involved in many activities in many countries.

It would be desirable, therefore, to have data for both establishments and firms. Unfortunately, reliable and comparable data classified by firm size are even rarer for small industry than data by establishment size. The CSO Census of Industrial Enterprises covers only those with 20 or more employees, which rules out much of the data in which we are interested. Consequently we are forced to rely mainly on establishment data. The position is not as damaging to our enquiry as might seem, however. For the vast majority of the units we will be examining in the study of small industry *per se*, the firm and the establishment come to the same thing. Usually, as Pryor (1972, p. 563) noted, "Establishments in multi-establishment enterprises are larger than establishments which are enterprises in themselves; further, the larger the enterprise, the larger the different establishments composing it." Table 1.1, which is based on UK data for 1980, confirms that picture. It suggests that few enterprises with less than 100 employees have more than one establishment and that the average size of establishment is an increasing function of enterprise size.

Table 1.1: *Relation between enterprises and establishments in UK manufacturing, 1980*

Enterprise size (in terms of numbers engaged)	*Number of enterprises*	*Average no. of establishments per enterprise*	*Average employment per establishment*
Less than 100	84,944	1.05	13
100– 199	2,437	1.50	93
200– 499	1,479	2.11	146
500– 999	599	3.82	183
1,000– 4,999	548	8.39	244
5,000– 9,999	79	24.84	281
10,000–19,999	48	33.36	418
20,000–49,999	21	58.38	553
50,000 and over	8	102.00	884
Total	90,163	1.20	56

Source: Department of Industry, *Business Monitor: Report on Census of Production 1980, Summary Tables,* PA 1002, Table 12, p. 234.

The Size Denominator

Whether we are dealing with enterprises or establishments, the question arises as to how they can most appropriately be classified by size. It would be possible to use some economic, legal or administrative concept, or some concrete variable such as output or employment. Any single basis of classification is necessarily limited, and these limitations should be borne in mind.

The Bolton Report (1971) took the view that the unique and common characteristic of small firms is that they are managed by the people who own them. This is certainly an interesting and useful way of segregating firms for various forms of analysis, but as a general denominator of size it would give rise to certain anomalies. For example, a firm owned and managed by one family might employ over 500 workers, whereas a public limited liability company with many shareholders and professional management might employ less than 200 workers. For some purposes, it might be more appropriate to regard the former as "large" and the latter as "small". There is also the problem of how to treat a workers' co-operative which may range from a small number of employees to a considerable number. Furthermore, there are major problems in getting the necessary data on ownership and control, so much so that in its quantitative analysis, the Bolton Report itself was forced to fall back on denominating firms as large and small in relation to their employment.

In fact, employment is by far the most commonly used size denominator in official statistics. It has many advantages. It is readily understood by the layman, and indeed accords with the way he would instinctively measure the size of firms. Although labour might be viewed as just another input into production, it is usually a substantial input, and one involving immense social significance. It is comparatively easy to collect data on numbers employed, and to compare these data over time and between countries. It becomes possible to define and measure *gradations* of size, whereas a conceptual denominator like that proposed by Bolton may have only a single cut-off distinguishing large from small.

Nevertheless, there are analytical difficulties attached to using the crude numbers employed as the size denominator. Although all men (and women) may be equal under God, in economic terms labour is by no means homogeneous. To classify as the same size two establishments, each with 100 workers, could give rise to obvious anomalies where one consisted largely of highly-skilled technologists, while the other consisted mainly of juveniles and unskilled workers. Moreover, labour as an input can be substituted for by other factors of production, and the degree of substitution will depend partly on relative factor prices which vary over time and between countries. Thus, the scale (in terms of other important variables such as output or capital employed) of a textile firm employing 100 workers today may be vastly different from one employing the same number fifty years ago. With the general secular tendency for labour productivity to rise, the use of labour, compared with either output or capital, as the size denominator, underestimates the rise in average size of establishments over time.

The use of capital employed as a denominator of scale suffers from the same limitation as employment in that it is only one of the inputs. But, in addition, the conceptual and practical problems attached to measuring capital employed[1] are vastly greater than in the case of labour.

In principle, the value of output would seem an ideal denominator since it may be viewed as the sum of the volume of inputs used, weighted by their explicit or implicit rates of return. Two possible measures of output suggest themselves, *gross output* and *net output*, the latter being a rough proxy for *value added*. Gross output is the selling value of production (adjusted for changes in stocks), and corresponds to what is commonly called turnover. A major problem about this measure, however, is that only a proportion of the value of turnover is produced within the firm itself — the rest being materials and services purchased from other firms — and this proportion is highly variable between industries, and even between firms in the same industry. Thus, a firm may be relatively large in terms of turnover but small in terms of most other key indicators of scale, such as value added, employment and fixed capital. Turnover may be important for some purposes — as a determinant of the amount of working capital, for instance — but unless the structure of production is broadly similar between firms, it cannot be regarded as a good general denominator of scale.

Value added is free from the objection mentioned in the previous paragraph, and since it represents the value of the firm's own activity, and is the fruit of all the factors it owns or controls, it would appear to be the most useful general denominator of size. There are some difficulties, however, which make it less satisfactory in practice than in principle. A firm's value added may be less stable as a denominator of size than other measures such as employment or turnover. That is because value added consists of wages and profits, and the latter may be either positive or negative in a particular year. Furthermore, for many countries, including Ireland, value added data are not generally available for all firms. Instead what is usually available is net output, which is the difference between the value of gross output and materials purchased; thus net output includes not only value added but also purchases of services such as printing, advertising, accounting and legal fees, etc.

Whether gross output or value added is used as the size denominator, there are difficulties in making comparisons between countries, or over time in the same country, because of changing prices. Often the only method available for converting the value figures for different countries to a common base is the foreign exchange rate, which is well known to be subject to serious limitations for this purpose. Over time within a country, there are general and relative price changes, and it is difficult to find suitable deflators at firm level to convert the data for different years to a constant price basis.

The foregoing suggests that there is no uniquely satisfactory single denomi-

[1]See Vaughan (1980) for a full discussion of these problems.

nator of size. In principle, this need not be a problem in multivariate analysis since different size variables can be used and attempts made to determine their separate effects if any. In practice, however, the different size variables, while not perfectly correlated, may be correlated to such an extent that it is impossible to distinguish their separate effects - the problem of multicollinearity. For the purpose of policy administration, it is generally not practicable to work with a whole range of size denominators, and simple and somewhat arbitrary denominators must be employed. In practice also, data on the desired size variables may simply not exist, and we must make do with what is available. Employment data are the most generally available, and that reason alone largely explains why employment is the most commonly-used size denominator.

The Dividing Line between Small and Large Manufacturing Industry

Assuming that size is denominated in terms of employment, the question remains as to where to place the dividing line between small and large manufacturing firms or establishments. Practice differs between countries, and even among different agencies within the same country. In Ireland the Industrial Development Authority (IDA) defines "small" as relating to establishments with up to 50 persons engaged, though there is an additional limit in terms of the amount of fixed asset investment (£500,000 in 1983). The Small Firms Association — a group established within the Confederation of Irish Industry (CII) to represent the interests of small firms — caters for firms employing up to 100 persons. In the UK, the Bolton Report (1971) measured as small those manufacturing firms or establishments with less than 200 persons engaged. The legal definition of small firms in Japanese manufacturing relates to those employing up to 300 persons (Bannock, 1976). In Austria, Belgium, France, Sweden and Switzerland, small firms are usually taken to mean those with less than 50 employees, while Germany and Itally normally include firms with up to 100 employees in that category (OECD, 1971a). In the US, the Small Business Administration (SBA) varies its definition for different purposes. For example, in determining access to SBA loans, the general upper limit applying in manufacturing is 250 employees, but in some industries the limit can go as high as 1,000, or even up to 1,500. In developing countries the dividing line is closer to that used by the IDA in Ireland, and certainly firms with more than 100 employees are not normally classified as small. The different concepts of small partly depend, of course, on the differences in industrial structure between countries. If Ireland were to classify as small all manufacturing establishments with less than 200 persons engaged, this would cover about half of total manufacturing employment, whereas in the UK it would cover only about one-quarter.

For analytical purposes it is not necessary to establish a single dividing line, and indeed it is often more useful to consider many ranges of size. For administrative purposes, however, it is convenient to have simple and consistent dividing lines. To accord with IDA practice in Ireland, we shall generally reserve the term "small" for establishments or firms with less than 50 persons engaged. We shall also be interested, however, in other ranges of size, though in many tables in order to simplify the presentation, we shall compress the ranges into four categories, labelled as follows:

	Persons Engaged
(1) Small	Less than 50
(2) Small-medium	50–199
(3) Medium-large	200–499
(4) Large	500 and over

It will be clear from the foregoing discussion that whereas only category (1) is treated as small here, in some of the developed countries category (2), and even category (3), would also be treated as small.

Data Problems in Relation to Small Industry

The data in relation to small manufacturing industry, particularly the very small firms, often leave much to be desired. Part of the problem is that there are so many very small firms, consisting of only one or a few people, that these firms are sometimes hard to locate, and that they often have difficulty in giving satisfactory statistical returns. For this reason, many countries adopt a cut-off point and exclude the very small firms from their industrial inquiries. The Irish Central Statistics Office, for instance, does not normally attempt to cover establishments with less than three persons engaged. It is clear, however, that some such establishments have entered the Census of Industrial Production (CIP), since the *average* employment in the establishments with less than five persons engaged was 2.83 in 1968 and 2.97 in 1975. Moreover, a significant number of small establishments with three or more persons engaged were not covered in practice in the CIP until 1979, when the coverage was significantly improved. Other countries may adopt a higher cut-off point, and some may adopt no cut-off point at all, so that cross country comparisons can be very difficult if not outright impossible in some cases. The comparisons are affected more in relation to establishment numbers than employment (or output, or capital). For example, in Ireland, in 1975, establishments recorded in the CIP with less than five persons amounted to 10.6 per cent of all establishments but only 0.5 per cent of CIP manufacturing employment. If in reality the true number of establishments with less than five persons engaged were, say, four times the recorded number, then such establishments as a proportion of all establishments would rise to 32.2 per cent, but their employment share would only rise to 1.6 per cent (assuming that the average employment per unrecorded establishment was two persons).

A further problem is where to draw the line between manufacturing and other economic activities, a problem which is often difficult in the case of small firms and on which practice differs between countries. For example, in Germany, repair work or craft-work carried out in retail establishments (such as baking, sausage-making, upholstery, silversmiths, shoe repairs etc.) is included in manufacturing, whereas in Ireland or the UK these would generally be included in the distribution sector or other service trades. An indication of the scale of the incomparabilities that can arise is given in the Bolton Report (1971), where in order to try to render the German data comparable with UK data, *Handwerk* activities were excluded from the German data. This involved reducing the recorded number of German manufacturing establishments by 331,000 — more than $3\frac{1}{2}$ times the *total* recorded number of UK manufacturing establishments! Another way of illustrating the difference, is that although the number of manufacturing establishments with 10 or more persons engaged is broadly similar in the two countries, the recorded number of manufacturing establishments smaller than this is ten times greater in Germany than in the UK. Although this enormous difference is partly a genuine indication of the greater prevalence of the manufacturing crafts in Germany, it is also due to different ways of classifying the same data.

As we shall see, there are other problems affecting the comparability of data on small manufacturing industry for different countries or different periods. Enough has been said, however, to show that caution must be exercised in comparing different data sets, particularly when they relate to different countries.

2 *Outline of Study*

Having discussed some of the major definitional and measurement problems that arise in studying small industry, we now outline briefly what this study contains. Part I examines the evidence available on the role of small manufacturing generally, with the prime emphasis on Ireland but drawing on data from other countries as well. It begins in Chapter 2 by examining data on the size structure of manufacturing industry and how this has changed over time. Chapter 3 considers how well small firms are adapted to carrying out the major business functions — technological capability, finance, marketing and overall management. In Chapter 4 we examine the structure and growth of the small firm sector in an attempt to answer questions as to why small firms continue to exist, in which activities they tend to be most highly concentrated, how their overall efficiency compares with larger firms, and what is their capacity for growth. Small industry is sometimes regarded as the seed bed of new entrepreneurship, and in Chapter 5 we take up the question as to what, if any, importance can be attached to the entrepreneurial function and the role of small industry in developing it. Chapter 6 outlines policy approaches in other countries to the development of small manufacturing industry, as a background to considering the approach in Ireland.

Part II of the study is devoted to an investigation of the IDA Small Industry Programme (SIP) which has been in operation since 1967. Chapter 7 describes the scheme in the context of general industrial development policy in Ireland as it relates to small firms. Chapter 8 examines the contribution of the SIP in terms of employment. Chapter 9 looks at the cost of the programme in terms of grants, and gives details of the amount of investment, the degree of capital intensity and the grant cost per job. Where possible, performance is assessed in comparison with the rest of industry and with other IDA programmes.

The final part of the study is concerned with the policy implications for Ireland. In considering future policy, however, the discussion cannot be confined simply to the lessons of the past, since industrial policy must be shaped with regard to the likely future overall economic environment, which in many respects will differ from the past. Moreover, future policy for small industry must properly be set in the context of general industrial policy, which is now in process of major overhaul.

Part I

THE ROLE OF SMALL MANUFACTURING INDUSTRY

Chapter 2

THE EXTENT OF SMALL MANUFACTURING INDUSTRY IN IRELAND AND ABROAD

In this chapter we examine the data for Ireland on the extent of small manufacturing industry relative to total manufacturing industry, and how it has changed over time. We compare the position in Ireland with a range of other countries for which data are available.

1 The Size Structure of Irish Manufacturing Industry

Two principal data sources are used to analyse the size structure of Irish manufacturing industry: CSO data from the Census of Industrial Production[2] and the IDA data from their Annual Employment Survey. A Census of Industrial Production is carried out each year, and an analysis of these data by size is published at intervals, the latest data relating to 1980. The IDA Annual Employment Survey started in 1973. Unfortunately, the two sets of data differ in important respects, and in particular the CSO data were less complete in relation to very small establishments prior to the extension of coverage in the 1979 CIP. Since the two sets of data cannot readily combined, we discuss them separately.

The Size Structure Based on CSO Data

The Annual Census of Industrial Production attempts to cover all manufacturing establishments except those with less than three persons engaged. Yet total manufacturing employment recorded in the CIP is significantly less than in the Census of Population, which is generally regarded as giving the most complete picture. In 1971, for instance, CIP manufacturing employment was 196.3

[2]The CSO Census of Industrial Enterprises is also available since 1975. Unfortunately, the data exclude all enterprises with less than 20 persons engaged, so that they are of limited use in a study of small industry as defined here.

thousand as against 213.6 thousand in the Census of Population.[3] This disparity was not due solely to the exclusion in the CIP of firms with less than three persons engaged. This is now confirmed by the extension of coverage in the 1979 CIP following the increased use by the CSO of administrative and public utility records, the employment of field officers, and the inclusion of estimates for establishments on the basis of limited information.[4] As a result, the number of establishments engaged in manufacturing in 1979 was increased by 683 (to 4,588) compared with the former coverage, while total manufacturing employment was raised by 10,000 to 228,000.[5] The vast majority of the additional establishments were small, with nearly three-fifths having less than 10 persons engaged (CSO, 1984).

A further point to be borne in mind is that the disparity between the CIP total for manufacturing employment and that in the Census of Population has narrowed over time — even before the large, explicit increase in coverage in 1979. In 1946, for example, the gap was 37 thousand (33.6 per cent of the CIP total), as against 17 thousand in 1971 (8.8 per cent). In part, this may reflect a genuine decline in the relative and absolute importance of one-person or two-person concerns. But it may also reflect improvement in the coverage of firms included in the CIP. Hence, the trends over time in relation to small industry as revealed in the CIP must be interpreted with a degree of caution throughout, and since 1979 they cannot validly be compared at all with the figures for earlier years.

Table 2.1 summarises the data on the number of establishments in various size classes in Irish manufacturing industry in the CIP for various years from 1929 to 1980, the former being the first year, and the latter the latest year, for which these data are available. Size is measured in terms of total persons employed.

The total number of establishments in CIP manufacturing was 50 per cent higher in 1975 than in 1929. The big increase came in the period 1931-38, when extensive trade protectionist measures were adopted with a view to developing Irish manufacturing. Part of the increase during the 1930s, however, may reflect

[3]It is true that outside piece-workers, numbering about 4,500 in 1971, are not included in the CIP total but probably appear in the Census of Population. This factor was more than offset, however, by the understatement of female employment in the Census of Population. (On this, see Walsh (1971), Appendix 1, pp. 111-113). Thus, the Census of Population itself did not give the full total of manufacturing employment, so that the extent of understatement of manufacturing employment in the CIP is greater than the above figures would suggest.

[4]Furthermore, two other activities, poultry processing and photographic laboratories, have been added which were formerly not included in the CIP.

[5]The 1979 CIP figure of 228,000 for total manufacturing employment compares with a Census of Population related estimate of 239,000 given in the *Economic Review and Outlook*, Summer 1983. The discrepancy is due, *inter alia*, to the exclusion of businesses with less than three persons engaged in the CIP.

Table 2.1: *Number of establishments in Irish manufacturing industry classified by size, various years, 1929–1980 (CSO data)*

<table>
<tr><th>Establishment size (in terms of numbers engaged)</th><th>1929</th><th>1931</th><th>1936</th><th>1938</th><th>1946</th><th>1958</th><th>1963</th><th>1968</th><th>1973</th><th>1975</th><th>1980 (extended coverage)[a]</th></tr>
<tr><td><20</td><td>1,546</td><td>1,558</td><td>1,999</td><td>2,135</td><td>1,769</td><td>1,808</td><td>1,649</td><td>1,512</td><td>1,508</td><td>1,693</td><td>2,878</td></tr>
<tr><td>20– 49</td><td>366</td><td>388</td><td>563</td><td>581</td><td>607</td><td>703</td><td>717</td><td>793</td><td>792</td><td>771</td><td>962</td></tr>
<tr><td>Total <50</td><td>1,919</td><td>1,946</td><td>2,562</td><td>2,716</td><td>2,376</td><td>2,511</td><td>2,366</td><td>2,305</td><td>2,300</td><td>2,464</td><td>3,840</td></tr>
<tr><td>50– 99</td><td>141</td><td>145</td><td>220</td><td>215</td><td>237</td><td>291</td><td>336</td><td>353</td><td>411</td><td>411</td><td>475</td></tr>
<tr><td>100–199</td><td rowspan="2">122</td><td rowspan="2">114</td><td rowspan="2">190</td><td rowspan="2">210</td><td>147</td><td>161</td><td>199</td><td>223</td><td>234</td><td>250</td><td>301</td></tr>
<tr><td>200–499</td><td>83</td><td>112</td><td>131</td><td>142</td><td>176</td><td>143</td><td>149</td></tr>
<tr><td>500+</td><td>11</td><td>9</td><td>19</td><td>15</td><td>21</td><td>31</td><td>45</td><td>51</td><td>59</td><td>52</td><td>55</td></tr>
<tr><td>Non-attributable</td><td>—</td><td>—</td><td>—</td><td>—</td><td>—</td><td>—</td><td>—</td><td>—</td><td>—</td><td>—</td><td>71</td></tr>
<tr><td>Total</td><td>2,186</td><td>2,214</td><td>2,991</td><td>3,156</td><td>2,864</td><td>3,106</td><td>3,077</td><td>3,074</td><td>3,180</td><td>3,320</td><td>4,891</td></tr>
</table>

[a]As explained in the text, the data for 1980 are based on an extended coverage and include several hundred small establishments not previously recorded. They cannot, therefore, be compared with the data for earlier years.

Source: Census of Industrial Production. Figures for 1946 and earlier years were adjusted for comparability with later years.

no more than an improvement in the coverage of the CIP, particularly in the case of the smaller establishments (Farley, 1975); but there is no doubt that there was also a major influx of new establishments in this period. During the war, there was a decline in the number of establishments to a low point of 2,800 in 1942, but there was a rapid increase in the immediate post-war years to a peak of 3,359 in 1952.[6] Subsequently, the numbers declined to 3,074 in 1968, but recovered again to 3,320 in 1975.

If we choose 50 persons engaged as the dividing line between small and large establishments, then the vast bulk of Irish establishments must be classified as small. In 1975, establishments with less than 50 persons engaged accounted for nearly three-quarters of all establishments.[7] This proportion is considerably lower than in 1929, however, when such establishments constituted 87½ per cent of all establishments. Even in absolute terms, the number of such establishments was declining from 1958 to 1973. This decline, however, was concentrated in the very small establishments with less than 20 persons engaged, the numbers of which fell in most years from 1938 to 1968. Moreover, within the category of less than 20 employees, the numbers of those with less than 10 employees fell much more than those with 10–19 employees. From 1973 to 1975, however, there seems to have been a reversal of the decline in numbers of small establishments. In part, this might reflect the fact that total employment declined in this period, pushing more establishments into the lower size category. As we shall see from the IDA data, however, the change may be more significant than this.

While small establishments dominate numerically, their importance in total employment is quite different, as may be seen from Table 2.2. Establishments with less than 50 persons engaged, amounting to three-quarters of all establishments, accounted for only one-fifth of total employment in 1975. Moreover, since we shall see later in Chapter 4 their level of productivity is below average, their share in output would be lower still. At the top end of the size range, establishments with 500 or more persons engaged accounted for a broadly similar share of total employment, although they represented less than two per cent of establishments numbers.

The relative importance of small establishments in terms of employment has changed considerably over time. In 1929, those with less than 50 persons engaged accounted for 34 per cent of manufacturing employment, but since then their share has declined. The decline has been greatest in the case of those with less than 20 employees, though there was a slight reversal of this trend between 1968 and 1975. The employment share of those with 20–49 employees rose slightly in the era of intensive protection in the 1930s, but subsequently has

[6]The data for 1942 and 1952 are not shown in the table since only an aggregate figure is available.

[7]In 1980, on the basis of the extended coverage of the CIP, such establishments accounted for nearly four-fifths of all establishments (78.5 per cent).

Table 2.2: *Distribution of employment in Irish manufacturing industry by establishment size, various years, 1929–1980 (CSO data)*

Establishment size (in terms of numbers engaged)	*1929*		*1938*		*1946*		*1958*		*1963*		*1968*		*1975*		*1980 extended coverage*[a]	
	No. '000	*% of Total*	*No. '000*	*% of Total*	*No. '000*	*% of Total*	*No. '000*	*% of Total*	*No. '000*	*% of Total*	*No. '000*	*% of Total*	*No. '000*	*% of Total*	*No. '000*	*% of Total*
<20	11.3	17.1	15.4	15.4	14.1	12.8	15.0	10.6	14.2	8.4	13.1	7.1	15.5	7.9	26.0	11.5
20– 49	11.0	16.6	17.4	17.4	18.4	16.7	22.5	15.9	22.7	13.5	25.1	13.6	24.3	12.5	30.3	13.4
Total	22.2	33.6	32.8	32.7	32.5	29.5	37.5	26.4	36.8	21.9	38.2	20.8	39.8	20.4	56.4	24.9
50– 99	9.0	13.7	13.8	13.8	15.2	13.8	20.6	14.5	23.1	13.7	24.6	13.4	29.0	14.9	33.3	14.7
100–199	23.0	34.8	39.6	39.5	19.5	17.7	21.7	15.3	27.7	16.5	31.9	17.3	34.8	17.9	42.2	18.6
200–499					23.9	21.7	32.4	22.8	38.8	23.1	42.1	22.9	43.1	22.1	44.6	19.7
500+	11.8	17.9	14.2	14.2	18.9	17.2	29.6	20.9	41.8	24.9	47.2	25.7	48.1	24.7	48.6	21.4
Non-attributable	—	—	—	—	—	—	—	—	—	—	—	—	—	—	1.7	0.7
Total	66.1	100	100.2	100	110.0	100	141.8	100	168.2	100	183.9	100	194.8	100	226.8	100

[a]See note to Table 2.1.

Source: Census of Industrial Production, for 1958, 1963, 1968, 1975 and 1980. The figures for 1938 and 1946 are taken from Linehan (1962). The figures for 1929 are estimates based on the data available for the number of establishments in each size range in these years: the average employment in each size range was taken to be the same as in 1938, except for the 500+ class which was derived residually as the difference between total manufacturing employment and the estimated total of employment in the other size ranges. As in Table 2.1, the figures for 1946 and earlier years were corrected for comparability with later years.

trended downwards. The employment share of establishments in the size ranges from 50 to 500 employees, which account for more than half of total manufacturing employment, has been remarkably stable since 1938. Thus the chief long-term change has been the gain in the share of establishments with more than 500 employees at the expense of those with less than 50 employees, though there is a hint of a decline in the share of the large establishments since 1968. Although the employment *share* of establishments with less than 50 employees has fallen substantially, the *absolute* numbers of workers was over 80 per cent greater in 1975 than in 1929. The big increase in numbers, however, came in the 1930s and after the Second World War, and from 1958 there was only a slight rise.

It does not necessarily follow from these data, of course, that larger establishments were growing more rapidly than smaller establishments. The establishments that appear in one size group in any year are not necessarily those that appear in the same size group at a later year. Some will have grown to a larger size range, some will have declined, some will have gone out of business altogether, and some new establishments will have entered. What can be said is that the outcome of these forces resulted in a strong tendency towards an increasing concentration of employment in larger establishments — at least up to the end of the nineteen sixties.

The Size Structure Based on IDA Data

The IDA have conducted an annual survey of employment in manufacturing at the beginning of each year since 1973. A small number of service-type establishments are included in the survey, but these can be identified and excluded, as is done here. The survey aims to cover all manufacturing establishments, and the industrial classification used is the same as that used by the CSO. The IDA admit, however, that their coverage of smaller establishments in the Dublin area may not be complete, because of the difficulty in identifying all such establishments in a large urbanised area. Prior to 1979, the data for such establishments was based on a sample survey, which was later considered to be defective. In 1979 complete census coverage was extended to Dublin, and estimates of employment were made for earlier years for those firms still surviving. Data were generally not available, however, in respect of firms not previously in the survey and which had closed by 1979. Accordingly there is probably some degree of overstatement in the trend of employment increase for small firms shown by these data.[8]

Total manufacturing employment recorded in the IDA survey in 1980

[8]O'Farrell (1984a) puts the employment level in 1973 in Dublin plants with less than 50 engaged, which were never recorded and had closed by 1979, at 8,300. There are some reasons, however, for regarding this estimate as too high.

amounted to 240,500. This total is very close to the estimate of 243,000 for April 1980 on the Census of Population basis given in the *Economic Review and Outlook*, Summer 1983. The concordance between the IDA figures and the estimates on a Census of Population basis also seems to hold for changes between 1973 and 1980. The IDA data show a rise from 217,600 in 1973 to 242,500 in 1980, while the population-based data show a rise in the same period from 220,000 to 243,000.[9] These increases are higher, however, than the CIP rise over the nearest similar dates available, estimated on the basis of the 1979 extended coverage, from 212.6 in the March quarter 1973 to 231.6 in the December quarter 1979.

Table 2.3 sets out the size distribution of Irish manufacturing industry based on IDA data for the years 1973 and 1980. If we first compare the 1980 IDA data with the CIP data for 1980 on the basis of the extended coverage, given in Table 2.1, it emerges that there were 637 more establishments in the former. The major difference was concentrated, however, in establishments with less than five persons engaged, where there were 820 more establishments in the IDA survey than in the CIP. No doubt, the bulk of these would have less than three persons engaged, which are deliberately not covered in the CIP. The IDA data in this regard are not comprehensive either, but small establishments (with less than 50 persons engaged) accounted for a rather higher proportion of the total number of establishments in the IDA survey as compared with the CIP — 80.5 per cent as against 78.5 per cent. In terms of employment, however, the distributions are not all that different. In the IDA 1980 survey, the small establishments accounted for 23.7 per cent of employment as compared with 24.9 per cent in the 1980 CIP. In the other size classes the employment shares are also very similar in the two data sets.

Looking at the changes in structure over the period 1973-80, as shown by the IDA data, the most significant change relates to the large establishment class, with 500 or more persons engaged, where employment fell both absolutely and as a percentage of the total. This decline may reflect forces special to the 1970s rather than a long-term trend. These forces include the impact of free trade on the older Irish import-substituting industries, the world depression following the oil crisis of late 1973, and factors peculiar to particular firms such as the closure of the Ferenka plant, employing about 1,500 persons, in late 1977.

As regards small industry (with less than 50 persons engaged) the data suggest some upsurge in the absolute number of such establishments, though their relative numbers continued to fall slightly. Employment in small establishments rose by 13.2 per cent from 1973 to 1980, somewhat greater than the overall rise of

[9]The 1973 figure is derived from Sexton (1982).

Table 2.3: *Number of establishments and employment in Irish manufacturing, classified by size, 1973 and 1980 (IDA data)*

Establishment size (in terms of numbers engaged)	*1973*				*1980*				*% change in employment 1973–80*
	Establishments		*Employment*		*Establishments*		*Employment*		
	No.	*% of Total*	*No. ('000)*	*% of Total*	*No.*	*% of Total*	*No. ('000)*	*% of Total*	
Small (<50)									
<5	1,241	25.7	3.0	1.4	1,438	26.0	3.3	1.4	11.7
5– 9	819	17.0	5.6	2.6	951	17.2	6.5	2.7	15.8
10–14	544	11.3	6.4	2.9	575	10.4	6.8	2.8	6.1
15–19	336	7.0	5.6	2.6	377	6.8	6.3	2.6	12.8
20–29	476	9.9	11.5	5.3	542	9.8	13.0	5.3	12.7
30–49	497	10.3	18.7	8.6	568	10.3	21.6	8.9	15.5
Total Small	3,913	81.0	50.7	23.3	4,451	80.5	57.4	23.7	13.2
Small-medium (50–199)									
50– 99	411	8.5	28.9	13.3	498	9.0	34.9	14.4	20.8
100–199	275	5.7	37.4	17.2	338	6.1	46.7	19.3	25.0
Total small-medium	686	14.2	66.2	30.4	836	15.1	81.6	33.7	23.2
Medium-large (200–499)	170	3.5	51.1	23.5	180	3.3	54.0	22.2	5.6
Large (500+)	60	1.2	49.7	22.8	61	1.1	49.5	20.4	-0.3
	4,829	100	217.6	100	5,528	100	242.5	100	11.4

Source: IDA *Annual Employment Survey.* Data relate to 1 January in the year concerned. As explained in the text, there may be some understatement in the 1973 figures for establishments with less than 50 persons — due to deficiencies in the Dublin data — and hence an overstatement of the rise in employment for 1973–80. Also, the data in relation to establishments with one or two persons engaged is thought to be collected erratically: however, this would not greatly affect the employment position given above because of the small aggregate employment involved in such firms.

11.4 per cent.[10,11] The chief gain in employment in this period was in the small-medium size range (50-199 persons), where employment rose by 23.2 per cent.

2 *Comparison with Other Countries*

The difficulties of making international comparisons regarding the relative importance of small industry are notorious, and the main problems have been outlined in Chapter 1. For the reasons given there, the problems are particularly acute as regards the numbers of firms or establishments. It is safe to say, however, that no matter which data classification is used, in all countries the great majority of establishments fall into the category of small — on any reasonable definition of small. But, precisely, because there are so many small establishments in every country, differences in definition and coverage can seriously affect comparisons of the size structure of industry across countries based on numbers of establishments. Because the establishments that give rise to comparability problems, however, usually have low employment, international comparisons of the size structure based on employment are less seriously affected by the differences in definition and coverage.

Table 2.4 gives data for a range of countries for the most recent available year on manufacturing employment divided into the four size classes of establishments specified in Chapter 1: small (1-49 persons engaged), small-medium (50-199); medium-large (200-499) and large (500+). The proportion of total manufacturing employment accounted for by small establishments varies considerably across countries from as low as nine per cent in West Germany,[12] to as high as 47 per cent in Japan due largely to the high degree of sub-contracting arrangements in that country. Ireland with 24 per cent comes about mid-way in the range, but, nevertheless, has a considerably lower share than some of the

[10]As explained earlier, there may be some overstatement in the rise in total employment, and more particularly in small industry employment, due to deficiencies in the Dublin data for small firms in 1973. Excluding Dublin, however, the increases for the rest of the country, for which the data are considered to be complete, were 17.6 per cent for small industry as against 23.0 per cent for all manufacturing. Both these figures are higher than for the country as a whole, since the figures show a rise in employment in Dublin of only 4.2 per cent in small industry and a fall of 7.2 per cent in all manufacturing. If as suggested by O'Farrell (1984a) the Dublin 1973 level of employment in small industry is understated, then the Dublin experience from 1973-80 would be even much worse than these figures show.

[11]The CIP data on the basis of the extended coverage available for 1979 and 1980 also suggest a relatively greater rise in employment in small industry, which rose between these two years by 5.8 per cent whereas overall manufacturing employment fell by 0.6 per cent.

[12]If *Handwerk* activities were included, however, the share of employment in small establishments in West Germany would be considerably higher.

other developed small economies like Switzerland (38 per cent), New Zealand (32½ per cent) and Norway (29 per cent). In developing countries the small industry share would be even higher. For instance, the data in Economic Commission for Latin America (1981), which relate to the 'sixties and early 'seventies, show that more than half of manufacturing employment in several Latin American countries was in establishments with less than 50 persons engaged.

There are theoretical arguments and some empirical evidence supporting the view that establishment size differences between countries are strongly and positively influenced by the size of the domestic market (Pryor, 1972). The data in Table 2.4, however, give the impression that Ireland, though having a very small domestic market, does not have a particularly large concentration of employment in small manufacturing industry. This impression is confirmed by the data in Table 2.5 which gives details of the average size of manufacturing establishments for the countries listed in Table 2.4. Because of the serious incomparabilities that arise in the case of the very small establishments, the averages are calculated, first excluding all establishments with less than 10 persons engaged, and then excluding those with less than 20 persons engaged.

Whether we consider the average for establishments with 10 or more employees, or for establishments with 20 or more employees, Ireland emerges with a relatively high average size of establishment for a country of its size. Ireland is the second smallest[13] of the 20 observations in Table 2.5 (as measured by total employment in manufacturing), but it ranks twelfth in terms of average size of establishment (with 10 or more employees). It is, of course, true that average size differs considerably among different branches of manufacturing, and these differences tend to be systematic across countries — in the sense that the rankings of establishment size in different industrial branches in each country tend to be correlated internationally. This raises the possibility that the differences among countries in the average size of establishment in manufacturing as a whole may be strongly influenced by differences in the branch composition of manufacturing production. Previous research suggests, however, that in practice differences in the structure of manufacturing among countries are not an important determinant of the differences in overall average size (Linehan 1962 and Pryor 1972).

What is most notable about the Irish situation, however, is the high concentration of manufacturing employment in middle-sized establishments, and the low average size of its large establishments. As may be seen from Table 2.4, Ireland has 56 per cent of its total manufacturing employment concentrated in the two middle-sized classes — small-medium and medium-large — and this is

[13]The smallest, Northern Ireland, is a province of the UK rather than a separate country, and in that perspective the Northern Ireland data fit better into the general pattern.

Table 2.4: *Manufacturing employment in various countries, divided into establishment size classes**

	Small		*Small–Medium*		*Medium–Large*		*Large*		*Total*
Country and year	*No. '000*	*% of Total*	*No. '000*	*% of Total*	*No. '000*	*% of Total*	*No. '000*	*% of Total*	*No. '000*
United States (1977)	2,824.7	15.3	5,207.2	28.1	2,886.2	15.6	7,597.1	41.0	18,515.1
Japan (1978)	5,866.4	46.9	4,073.2	32.6			2,569.4	20.5	12,509.1
West Germany (1980)	678.1	9.1	1,592.8	21.3	1,379.4	18.4	3,831.8	51.2	7,482.0
United Kingdom (1979)	938.0	13.5	1,126.9	16.3	1,117.4	16.1	3,743.4	54.1	6,925.6
Italy (1971)	1,041.1	26.1	1,071.1	26.8	655.6	16.4	1,222.2	30.6	3,990.1
Spain (1978)	1,068.8	39.2	608.4	22.3	416.2	15.3	634.4	23.3	2,727.8
Canada (1975)	306.6	17.3	473.6	27.2	362.8	20.8	605.5	34.8	1,748.5
South Africa (1976)	167.0	12.3	311.6	22.9	302.2	22.2	579.2	42.6	1,359.9
Switzerland (1975)	366.5	38.4	241.4	25.3	138.7	14.5	207.1	21.7	953.7
Netherlands (1981)	154.5	18.0	203.0	23.7	135.4	15.8	363.5	42.5	856.3
Sweden (1981)	136.4	16.5	200.9	24.3	156.6	18.9	332.4	40.2	826.3
Belgium (1982)	185.7	23.5	170.6	21.6	134.8	17.0	299.8	37.9	790.7
Austria (1981)	68.6	11.5	281.8	47.3			245.7	41.2	596.1
Finland (1980)	89.9	17.2	148.6	28.3	121.8	23.2	165.0	31.3	525.3
Denmark (1981)	90.2	25.1	106.2	29.6	71.7	20.0	90.9	25.3	359.0
Norway (1980)	101.9	28.6	113.4	31.9	86.2	18.6	74.8	21.0	356.3
New Zealand (1975/76)	93.2	32.5	82.6	28.8	45.0	15.7	66.1	23.0	286.9
Israel (1980)	76.1	27.6	199.7	72.4					275.8
Ireland (1980)	57.4	23.7	81.6	33.6	54.0	22.2	49.5	20.4	242.5
Northern Ireland (1979)	15.6	12.0	33.4	25.7	34.2	26.4	46.5	35.9	129.7

*The size classes are as follows with the exception of the US: *Small*, less than 50 persons engaged; *Small-medium*, 50-199; *Medium-large*, 200-499; *Large*, 500 or more. For the US, the small-medium class covers 50-249 persons engaged, and the medium-large class 250-499.

Sources:

United States: *Census of Manufacturers 1977*, *General Summary*, Table 4.

Japan: *Japan Statistical Yearbook 1981*, Table 61.

West Germany: *Statistisches Jahrbuch 1982*. Data relate to "betriebe".

United Kingdom: "Report on the Census of Production 1979, Summary Tables", *Business Monitor*, PA 1002, Table 6.

Italy: *Annuario Statistico Italiano 1980*, Table 157. Data relate to "unita locali" and exclude 536,131 units (1,311,800 workers) of "artigiane".

Spain: *Censo Industrial de España 1978, Establiciemtos Industriales, Resumen Nacional*. Data include mining.

Canada: *Manufacturing Industries of Canada: Types of Organisation and Size of Establishments, 1975.*

South Africa: *South African Statistics, 1982.*

Switzerland: *Statistisches Jahrbuch 1982*, p. 151.

Netherlands: *Statistical Yearbook of the Netherlands 1982*, p. 186. Data cover only establishments with 10 or more persons engaged.

Sweden: *Industri 1981. Prelimmara Branschdata*. Data cover mostly only establishments with 5 or more persons engaged.

Belgium: *Rapport Annuel d'Office National de Sécurité Sociale*, Tables 15-16. Data include mining as well as manufacturing.

Austria: *Statistisches Handbuch 1982.*

Finland: *Yearbook of Nordic Statistics 1982*, Table 72. Data relate to establishments with five or more persons engaged.

Denmark: *Industri ög Energi 1983*, Statistiske Effterretminger. Data relate to establishments with six or more persons engaged.

Norway: *Yearbook of Nordic Statistics 1982*, Table 72. Data relate to establishments with five or more persons engaged.

New Zealand: *Census of Manufacturing Series A, General Statistics Bulletin No. 1, 1974-75 and 1975-76.*

Israel: *Statistical Abstract of Israel 1982*. Data include mining as well as manufacturing, but the mining component is very small.

Ireland: IDA *Annual Employment Survey.*

Northern Ireland: *Analyses of United Kingdom Manufacturing (Local) Units by Employment Size, Business Monitor*, PA 1003, Table 3.

Table 2.5: *Average size of establishments in manufacturing, various countries*

	Establishments with 10 or more persons engaged	*Establishments with 20 or more persons engaged*	*Establishments with 500 or more persons engaged*
	(1)	*(2)*	*(3)*
United States (1977)	111	152	1,324
Japan (1978)	47	93[a]	1,231
West Germany (1980)	n.a.	175	1,546
United Kingdom (1979)	140	210	1,599
Italy (1971)	58	99	1,364
Spain (1978)	56	89	1,201
Canada (1975)	97	133	1,305
South Africa (1976)	123	164	1,194
Switzerland (1975)	58	96	1,204
Netherlands (1981)	93	137	1,652
Sweden (1981)	101	147	1,298
Belgium (1982)	87	131	1,321
Austria (1981)	126	156	1,328
Finland (1980)	92	126	954
Denmark (1981)	67	101	1,096
Norway (1980)	65	98	947
New Zealand (1975–76)	59	90	918
Israel (1980)	80	131	n.a.
Ireland (1980)	74	100	812
Northern Ireland (1979)	122[b]	154	1,329

[a]Figure relates to 1975.
[b]11+.
Sources: As for Table 2.4.

higher than in any other country listed. On the other hand, Ireland ranks lowest in terms of the proportion of its manufacturing employment located in the large establishments (500+). This point is reinforced by Column (3) of Table 2.5 which shows that the average size of large establishments in Ireland is by far the lowest of any of the countries. If we compare Japan or Switzerland, for instance, which have nearly as small a share of their employment in large establishments, the average size of their large establishments is about 50 per cent greater than in Ireland.

In sum, while Irish manufacturing does not have a particularly high share of employment in small establishments or a particularly low overall average size of establishment, it also has a relatively low share of its employment in large establishments and their average size is small.[14] More so than in most countries, em-

[14]The EEC data for 1979 confirm that for *enterprises* also, the share of industrial employment in large enterprises and the average size of the large enterprises is low in Ireland.

ployment is concentrated in middle-sized establishments (from 50 to 500 workers).

Changes over Time in Other Countries

The evidence on long-term trends over time in the relative extent of small industry in different countries is not at all clear-cut. Conflicting findings have emerged from different studies. These conflicts often depend on differences in the time periods or in the range of countries considered. They also depend on differences in the size denominator used, in the dividing line between large and small, and on whether the basic entity considered is the enterprise or the establishment. These differences, together with the many incomparabilities that exist between countries in the coverage of the data, make it difficult to draw clear-cut conclusions.

As regards establishments, the International Labour Organisation (ILO) (1956) study for ten countries showed little change in the share of total employment accounted for by firms with 10 to 50 employees[15] in the thirty years or so up to 1950. The absolute numbers of employees rose in almost all countries, the only exceptions being the UK and France where total manufacturing employment was static or declining. Though there was a rise in the average size of establishment with ten or more employees in the US and in the UK, this tendency was not general: there was little change for the other countries, except Japan where there was a pronounced fall. At the top end of the size scale, however, almost all countries experienced a rise in the proportion of employees in establishments with 1,000 or more employees, the increases being particularly marked in the US, UK and Sweden.

Pryor (1972) presented results for six countries over a longer period (about 1910 to 1960), which differed somewhat from those of the ILO (1956), and from other studies such as Jewkes (1952) who also found comparatively little change in establishment size. Pryor rightly points out that the simple arithmetic average size is highly sensitive to the large number of very small establishments. To circumvent this problem, Pryor confined his analysis to establishments with 20 or more persons engaged. Furthermore, he used additional indicators: (1) the Niehans index which weights different establishment sizes not by the number of such establishments but by their share in total employment, and so attaches relatively high weights to the larger establishment sizes; and (2) the percentage of employment in establishments with 1,000 or more employed. All the indicators used pointed to a substantial increase in establishment size in the first half of the twentieth century. There was also some tendency in almost all of the countries for a shift in production towards those industries with relatively larger establish-

[15] Establishments with less than 10 employees were generally omitted because of doubts about the quality of the data.

ments, though the major influence was the general rise within industries.

While Pryor's results provide the most rigorously elaborated picture of the overall pattern of change, they are less helpful in regard to the position of small establishments. The exclusion of establishments with less than 20 employees removes from consideration a great part of the focus of attention in regard to small industry. Even apart from this, however, it is quite possible that average size of establishment would be rising and a larger proportion concentrated in very large establishments — without there necessarily being any diminution in the share concentrated in the small establishments. Pryor gives no data on the latter measure, which is probably the most useful indicator of the importance of small industry. Indeed apart from the ILO study, there is very little available in published studies on the trends in this measure in the first half of the twentieth century.

Since the 1950s there is more evidence on the trend of the share of small industry, but it does not present a simple pattern either. The OECD (1971a) study, which considered the evidence then available, concluded that in most countries the share of small industry was declining, whereas in a few, such as the US, it was maintaining its position. The data in the Bolton Report (1971) showed that in the decade from about the mid-1950s to the mid-1960s, the employment share of establishments with less than 200 workers fell in eight of the ten countries considered, while in the other two, the US and Canada, the share rose. More recently, Beckler (1980) notes that in the first half of the 1970s, small-medium establishments after first experiencing a relatively greater setback in the depression of 1974, recovered by 1976 to increase their share of total employment by one to three percentage points. More recently still, Binks and Coyne (1983) show that the share of establishments with less than 200 workers in total UK manufacturing employment rose between 1975 and 1979 from 28.9 per cent to 29.8 per cent.[16] These findings for the 1970s could simply reflect the shedding of employment in larger firms so that they now come into the small-medium class. It does not necessarily mean that small-medium firms were doing better: an equi-proportionate employment decline in firms at all size levels tends automatically to produce a larger proportion in the small size class. There have been, however, some developments in the 1970s favourable to small scale production, which are discussed in the next chapter.

All of the foregoing relates to establishments rather than enterprises. It is extremely difficult to find information for a range of countries on the changing size structure of enterprises over time. In the case of several European countries which give what purport to be enterprise data, what is usually involved is legal

[16]Not all countries shared this experience, however. Bollard (1984) notes that in New Zealand the number of small manufacturing firms has decreased in recent years, the first time this has happened there.

units — so that subsidiaries with an independent legal status are not grouped with the parent company. Common observation would suggest, however, that if the share of large establishments has been rising, then the share of large enterprises must be rising a great deal more, because of the tendencies towards mergers and take-overs. Indeed for the UK, Prais (1976) has demonstrated that the share in net output of the 100 largest manufacturing enterprises there has risen from about 16 per cent in 1909 to almost 41 per cent in 1970. In regard to manufacturing enterprises in the UK with less than 200 workers, the Bolton Report (1971, p. 58) showed that the decline in their share in total employment was much greater than the decline in the share of establishments with under 200 workers.

It does not necessarily follow, however, that the same would be true of very small enterprises. If we are talking about enterprises with less than 50 workers, then most such enterprises would have only one establishment. Moreover, probably only a limited number of establishments with less than 50 workers would be part of a larger enterprise. As suggested in Chapter 1, when we are dealing with a size of less than 50 employees, the concepts of "enterprise" and "establishment" largely overlap in practice, so that there may not be such a great difference in the trend over time in the relative share of this size category, whichever concept is used.

Pryor (1972, p. 557) aptly points out that "if an output measure had been used as a measure of establishment size, the increase in size would have been much greater ... because output per worker increased greatly." Again, however, it does not follow that the share of the output in small firms would have declined more (or increased less, as the case may be) than its employment share. That depends on the pace of productivity growth in small firms compared with large firms, an issue we consider later in Chapter 4.

3 *Conclusions*

As in all countries, small-scale establishments, defined as those with less than 50 persons engaged, represent the vast bulk of manufacturing establishments in Ireland, where they account for at least three-quarters of the total. Their share in employment, however, is less than one-quarter of the total and their share in output lower still. The employment share of small manufacturing establishments in Ireland is higher than in large countries such as the United States, and the United Kingdom, but is low by reference to many other small countries, such as Switzerland, New Zealand and Norway. At the other end of the size distribution, Ireland has a low share of employment in large establishments, and the average size of its large establishments is low by international standards — even by reference to other small countries at a similar level of development, such as

Finland. More than in most countries, therefore, manufacturing employment in Ireland is bunched in middle-sized establishments (50–500 employees).

The trends over time have, if anything, served to reinforce this dominance of the middle-sized establishments in Ireland. The longer-term trend in the employment share of small establishments was downwards in Ireland up to about the early 'seventies, since when there appears to have been some reversal of this trend — as in some other countries. While the employment share of large establishments in Ireland rose considerably in the first twenty years or so after the Second World War, it appears to have been falling since the late 'sixties.

Given this picture, it is understandable why, in the debate on industrial policy in Ireland, there has been growing emphasis on the need to build stronger and larger companies in Ireland — and we shall return to this topic in discussing policies in Chapter 10. But it should also be kept in mind that the evidence for other countries suggests that the presence of stronger companies is not incompatible with having a larger small firm segment, employing a significant proportion of the labour force in manufacturing. Even as it is at present, the proportion of total Irish manufacturing employment engaged in small industry is sufficiently great to justify attention to the special features of such establishments, particularly if some have potential for significant further development.

Chapter 3

SMALL-SCALE INDUSTRY AND THE MAJOR BUSINESS FUNCTIONS

The competitive success of any manufacturing business is crucially affected by its performance in relation to a number of major business functions: technological usage, marketing, financial control, and management. Size of firm often crucially affects, both positively and negatively, the ability of firms to carry out these functions satisfactorily. We now consider the evidence on the advantages and disadvantages of small firms in performing each of these major activities. The overall impact of scale on the economics of production is considered further in Chapter 4.

1 Technology

Technological change can be viewed as comprising the activities of research, invention and innovation. Research and invention are related to the development of technological knowledge, while innovation is the practical application of this knowledge in the form of new products or processes of production. Invention is nowadays often based on R & D but may also arise from empirical practice and experience inside or outside firms without any recourse to formal R & D. The first commercial use of an invention often involves large development expenditures on technical and marketing pre-tests. Technological knowledge is diffused internationally through trade in capital goods, direct foreign investment, licensing agreements, and the international movement of information and people. External economies in the development of new technology make for a high concentration of innovation in a limited number of locations. Firms, industries or countries which are at a distance from these innovation frontiers are, in many cases, imitators of advanced technology. Even then, application of new innovative ideas through technology transfer may still require a certain commitment to R & D on the part of imitators in order to identify opportunities and assimilate technological knowledge and skills.

Impact of Scale on Technology

The role of small firms in technological change has been the subject of much discussion in recent decades. A widely-held view up to about a decade ago was that small firms contributed little to technological knowledge or innovation due to increasing economies of scale in the development of technology. The limited resources and managerial expertise of small firms was thought to be insufficient to sustain an effective R & D programme, especially one employing highly qualified technicians, engineers or scientists, and using expensive equipment. The costs of developing new products up to the point of commercialisation and successful marketing tend to be particularly high and there may be no guarantee of "breaking even" in the initial years following the launch of the product. Moreover, unless competitors could be precluded (at least temporarily) from cashing in on research findings, the incentive to undertake R & D would be greatly diminished. Schumpeter (1950) was the most influential advocate of the view that large scale firms, enjoying a temporary monopoly, were the most powerful engines of technological advance.

An alternative view of the contribution of small firms has developed in recent decades especially in the US. It is argued that small firms may be particularly innovative in certain product sectors where economies of scale are not present or crucial to competition. It is claimed that small scale allows greater flexibility in responding quickly to market needs and new technological stimuli. The smaller the organisation, the generally less formal and bureaucratic is the management and running of an enterprise. Hence "the absence of organisational barriers facilitates communication and decisions are taken rapidly while the strategy is easily understood by all" (OECD, 1982, p. 31). Moreover, the individual owner/manager of a small enterprise may be highly motivated to experiment, innovate, and take risks, whereas in a larger organisation, management may be more conservative and also more security conscious. Small firms because of their vulnerability to market changes may be forced to engage in a process of continual experimentation in order to be able to survive. They operate at the margin of the market, entering and leaving more freely than larger firms: as such they are often the purveyors of dynamic industrial change.

In considering the empirical evidence underlying these views a note of caution needs to be introduced. As the evidence itself suggests, the role of small firms in contributing to technological change is changing over time; it varies considerably among different countries; it is related to industry sector and stage of product development; and it varies according to how radical is the innovation. Hence, one study for one particular country or one particular set of industries alone may be misleading as to the overall picture.

Furthermore, it is difficult to measure technological change directly, and different proxy measures can yield conflicting results. For example, while most of the studies in the last 20 years have concluded that inventive activity declines with size of firm — for a review of the evidence see Kamien and Schwartz (1975) — a recent study by Soete (1979) used a different measure of inventive activity, as well as a more finely graded breakdown of large firms, and concluded that some industries studied by him recorded increasing returns to inventive activity while others recorded decreasing returns. Finally, it should be noted that in many of the studies the smallest sized firms distinguished are large by reference to the size of small firms with which we are concerned. Proxy measures of inventive activity, such as the amount of R & D, are often of little relevance to small firms, since "the vast majority of small firms do not have sustained R & D programmes" (Kamien and Schwartz, 1975, p. 18).

The Bolton Report (1971, p. 50) concluded that "On the question of invention, as distinct from innovation, the evidence suggests that individuals working either by themselves or in small firms make a disproportionately large contribution." This observation is not altogether surprising in view of the relatively lower costs involved at this early stage of technological change, especially in the case of inventions that do not involve a high degree of scientific or engineering know-how. The later stages involve heavier capital and labour costs in testing and developing new products and processes. Jewkes, Sawers and Stillerman (1969) found that among the 70 most important twentieth-century inventions, more than half were made by individuals working on their own or with private help. Hamberg (1964) found that two-thirds of the major inventions in the period 1946–65 resulted from the work of independent inventors and small companies.

A somewhat different scenario emerges, however, when the empirical evidence for *innovation* is considered. Freeman (1971), in a survey of over 1,000 important innovations in the UK for the period 1950–70 covering manufacturing as well as construction and public utilities, found that small firms (those employing up to 200 persons) accounted for approximately 10 per cent of total innovations. This share of innovation was about half their share of employment and output but was more than twice their estimated share of research and development expenditure. The contribution of small firms to innovation was found to be lowest in the capital-intensive industries. The results were not significantly altered when the relative importance of innovations was considered. A survey based on a sample containing 319 major innovations in the US over the period 1953–73 showed that 23 per cent occurred in firms employing less than 100 (National Science Foundation, 1977). Small firms (those with less than 100 workers) contributed as much to innovation as their share of total employment. From this study, it was found that productivity in innovation, as measured by the number of major innovations per R & D dollar, tended to decline with size. Firms with less than 1,000 workers produced 24 times as many major innovations per R & D dollar as very large firms (i.e., over 10,000 employees) and four times as many as firms with 1,000–10,000 workers. In Japan, it was found that in the 1970s there was still a large disparity in the ratio of R & D expenditure to sales between small and large businesses. When a comparison was made by sector, however, it was found that the disparity was considerably narrowed. In particular small businesses were found to pursue R & D more positively than large in research-intensive activities in the chemicals, communications, electronics and electrical measuring instruments industries. (Ministry of Foreign Affairs, Japan, 1982).

It should be stressed of course that the definition and measurement of innovation is problematic, particularly for comparative purposes. Notwithstanding differences in the scope and measurement of innovation in the studies referred to above, however, it appears that smaller firms may be more innovative in the US compared to their counterparts elsewhere. Evidence for this is also provided by the work of Rothwell and Zegveld (cited in OECD 1982) who classified innovations according to the size of company sales. Small firms (those with a turnover of less than US$5 million per annum) made a relatively greater contribution to innovation in the US than in any of the other countries studied which included the UK, Japan, France and Germany. Furthermore, it was shown that small firms made a very significant contribution to radical technological breakthrough in the US compared to other countries. In fact, in the US, it was found that small firms were more significant in radical technological breakthroughs than they were in minor technological changes, a surprising result in view of the characteristically low contribution of small firms to radical change in other countries except in cases of the new embryonic technologies (micro-processors and biotechnology, for example). These findings suggest that the US may be exceptional in regard to the technological contribution of new small technology-based firms. Apart from the general socio-cultural environment and attitudes towards enterprise, there are several other favourable environmental factors: access to a large market with a relatively homogeneous legal/administrative system, sophisticated customer firms, ready information flows, and availability of specialised back-up services.

The contribution of small firms to technical change varies considerably with the type of industry sector. There are major sectors where small firms are of no consequence, such as aerospace, motor vehicles, many chemicals and cement; while their role is greater in activities such as scientific instruments, electronics, textiles and timber (Freeman, 1971; 1982). Economics of scale are the major explanation of why small firms contribute little to innovation in certain industries, but economies of scale vary over the life cycle of a product. Innovative small firms or plants tend to be concentrated at the early stage of the product cycle when technological development is rapid and fluid and the key to competitive success lies in non-price factors such as functional product performance.

(Abernathy and Utterback, 1976). In these firms, technology is less specific, equipment is of more general use, and manual labour and craft skills are important. In product sectors which have matured, technology becomes more specific, and competitive success is then more crucially dependent on price factors. Products become increasingly standardised and mass production is more efficient. In these product sectors, larger firms are the main contributors to innovation, which mainly takes the form of incremental improvements in the process of production rather than product innovation.[17] According to Utterback's (1979) findings based on historical studies of innovations in their organisational, technical and economic context, innovation in small firms is stimulated by producers' or consumers' needs and users' technical inputs, whereas in larger firms the stimulation to innovate arises from market pressures to reduce cost and improve quality. Shinohara (1968b) noted that technological progress in Japan was greater among small firms in producer than in consumer goods, in export activities than in domestic activities, in mechanical industry than in handicraft, and in cities than in rural districts.

In a discussion of the contribution of different size classes of firms to innovation it is important to stress the complementary roles of large and small firms, a point emphasised in the Bolton Report (1971) and by Storey (1982). In the UK, a very large number of innovations applied in large firms were the result of inventions and ideas originating in universities, government laboratories, small firms and the work of private inventors. Technology has been disseminated in both directions between large and small firms. Managers and employees working for large firms have in some cases resigned their posts and set up new innovative businesses themselves. In many cases licensing arrangements exist between a large and small firm supplying a market which would not be profitable for the large firm to supply. In Japan, many large corporations have small independent production units contained within their own major plants, which act as skilled sub-suppliers of components and inputs. In some cases, these sub-suppliers are former employees of the large corporation who set up business as an alternative to being made redundant. Strong linkages exist between large and small firms in both the US and Japan and this factor provides a major stimulus to innovation by small firms.

[17]Vernon (1966) also emphasised that many innovations occur early in the life-cycle of products, when firms in the industry are often relatively small. He also suggested that a disproportionately large number of new industries/products were developed in the US, the technological leader. This may be part of the explanation of the greater concentration of innovation in small firms in the US than in Europe: when European firms move into these industries a little later, the optimum scale of firm has increased due to greater standardisation, so that the more innovative sectors in Europe would be characterised by larger firm size. This still does not explain, however, why the US maintains the predominant leadership role in new technology development.

The fact that small firms have made a significant contribution to technological progress in various areas should not be taken to apply, however, to the generality of small firms. What is true for some is quite emphatically not true for all. On the contrary, the bulk of small firms in most countries tend to be technologically unsophisticated. As we shall see in later sections of this chapter, there are financial and managerial factors that pose difficulties for small firms in keeping abreast of technological developments.

We have been considering the relationship between size of firm and technological change primarily in terms of the impact of the former on the latter. But, in turn technological change affects the optimum scale of production, and this issue is considered further in the next chapter.

The Evidence for Ireland

The available evidence on small firm innovation in Ireland is limited, but the picture for indigenous manufacturing of all sizes is not favourable and is probably even less so among small firms. Several studies have reported low levels of technical innovation in indigenous industry in general and heavy reliance on foreign technology sources (Cooper and Whelan, 1973; Kieran, 1975; Allen, 1979; Telesis Report, 1982 and O'Brien, 1985).[18,19] Moreover, the level and efficiency of technology transfer within Irish industry was described as poor by Allen and Reilly (1973). To some extent, the reliance on foreign sources explains the low level of spending on research and development by firms in Ireland. In the last few years, however, the Industrial Development Authority has laid more emphasis on R & D, and grants for product and process development rose from 1.5 per cent of total IDA grant approvals in 1978 to 7.3 per cent in 1983.

A NBST survey of over 90, mostly small, firms in the engineering, chemicals and plastics sectors showed that only one-quarter of firms included in the sample could be regarded as innovative (NBST, 1982). The measure of innovativeness used showed declining innovation with decreasing employment size. The NBST study concluded that the technical capacity for either carrying out product and process development within firms, or for absorbing external assistance from public agencies or private industry, was low. Allen (1979) found that the firms he surveyed relied very little on universities or research institutes for developing new products or processes. A study of electronic firms in Ireland by Cogan and

[18]Exceptions, however, were Higgins (1977) and Murray (1983) which reported a more optimistic picture for particular industrial sectors, the former for food and the latter for engineering.

[19]In recent years the National Board for Science and Technology (NBST) has also carried out several studies of the technological condition of Irish industry, in particular indigenous industry. The NBST (1983a) *Science Budget 1983* gives the Board's views on the need for an integrated approach to technology and industrial policy in Ireland.

Onyenadum (1981) found a very low rate of company spin-off in the sector — only six spin-offs out of a total of 109 firms.

Technology transfer seems to occur mainly through direct foreign investment rather than licensing arrangements. According to the US Commerce Department (1981), the ratio of licence payments (from Ireland to the US) to US direct investment in Ireland was only two per cent in 1979 compared to four per cent for all countries. The number of successful joint ventures in licensing technology involving Irish and overseas firms is limited. The Industrial Development Authority, however, has introduced a technology transfer programme with a staff of seven officers abroad concentrating specifically on this activity.

The percentage of firms employing less than fifty workers which performed R & D was only two per cent in 1979 compared to 21 per cent of firms employing 100 or more (NBST, 1981). In 1982, firms with less than fifty employees accounted for only 5½ per cent of total R & D expenditure, very much below their share in total output or employment (NBST, 1983b). There has, however, been no comprehensive study of innovation, as distinct from R & D, in different sizes of firms in Ireland. The available studies have tended to focus only on a narrow sample of small firms and have not provided sufficient evidence to draw general conclusions. There is little doubt, however, that very few small Irish firms have been responsible for radical technological breakthroughs. Where new products have been introduced, it has usually been the result of a spin-off from a larger technologically-sophisticated firm or the direct copying of the technology of imported products.

2 *Marketing*

The Bolton Report (1971) attempted to classify small firms into three categories — satellites, specialists and marketeers — according to the type of market they supplied. The *satellites* depended heavily on one large customer and were found to be common in sub-supply activities in the motor, aircraft and engineering industries, as well as in the production of fashion goods and furnishing accessories to department stores. The *specialist* firms carried out functions that the large firms did not find it economic to perform, such as jobbing engineering. The third category of small firm, the *marketeers*, competed in the same or similar markets as large firms. The evidence suggested in fact that the vast bulk of small firms were marketeers — some 78 per cent of manufacturing firms were so classified in Merrett Cyriax Associates (1971). A significant proportion of the marketeers in manufacturing, however, were heavily dependent on one large customer, though not to the extent that they would be classified as satellites: some 35 per cent of all small firms in manufacturing were dependent on one customer for a quarter or more of their business.

Small firms frequently cater for markets which are locally based or specialised in character. This may arise from the personalised nature of the product or from the existence of significant transport and communications costs which inhibit competition against small firms supplying the local market or using locally supplied raw materials. The personalised service which often characterises small firms can be an important consideration especially in product lines requiring a high degree of specialisation and after-delivery service. Non-price factors, such as product design and quality, feature as important determinants in defending the competitive advantage of small firms. However, certain product markets requiring large outlays of expenditure on market research, advertising, distribution and transport may be inaccessible to small firms. Barriers to marketing for small firms may also exist due to tendencies towards oligopoly or monopoly in many product areas.

Historically, the lowering of transport and communications costs has tended to undermine the competitive position of many small firms relying on logistic cost advantages. Bollard (1983) has pointed out that the greatly increased cost of energy in the 1970s may have stimulated more decentralised production in some activities and thereby improved the market prospects facing small firms in these sectors.

Trade across national frontiers often necessitates elaborate distribution arrangements as well as a certain outlay on market research, advertising and promotion, beyond the resources of small firms. Added to these, poor managerial capacity and expertise may deter most small firms from exporting. Small firms typically need to develop experience first on the home market where familiarity, ease of access and transport costs facilitate the establishment of a competitive advantage. This presupposes the existence of an extensive local or national home market, which is not the case for many products in Ireland. The Telesis Report (1982) and Bradley (1983) noted a marked absence of an orientation towards exporting among indigenous manufacturing industry generally in Ireland. It would appear that many small firms in Ireland lack the financial and marketing expertise to launch into export markets.[20] In Table 3.1 data are given for the degree of exporting in each size category of industry in Ireland in 1973. Among firms with less than 100 persons engaged, almost one in every two had no exports at all as compared with one in five among firms with 500 or more employed. O'Farrell and Crouchley (1984) noted that the highest rates of new

[20]Export Credit Insurance and Finance Schemes are operated, on the basis of commercial assessment, by the commercial banks and other financial institutions. The schemes are underwritten by the government. The Government White Paper (1984) stated that steps were in train to improve these services, but was not very specific on details. The consultants' report in NESC (1984) stated that there was widespread ignorance about these schemes among exporters, and that small firms had made little use of them.

Table 3.1: *Percentage of output exported for different size categories of firms in Irish manufacturing industry, 1973*

Firm size (persons engaged)	*% of output exported*					
	None	*25% or less*	*25–50%*	*50% or more*	*Not stated*	*Total*
	% of firms in each export category					
Large (500+)	19	33	11	32	4	100
Medium (100–499)	22	27	10	27	13	100
Small (25–99)	47	25	7	14	7	100

Source: Gorman, Handy, Moynihan and Murphy (1974)

indigenous firm formation in Irish manufacturing from 1973 to 1981 occurred in non-traded activities largely protected from international competition.

Nevertheless, it would be unwise to assume that no small firms have export capabilities. A survey conducted in 1979, reported in Coras Trachtala (1980), suggested that while a much lower proportion of small firms were engaged in exports, the relative importance of exports among those that did export was at least as great as in larger firms. Taking those data in conjunction with the IDA data on the size structure of manufacturing, it would appear that over two-thirds of firms with less than 50 employees had no exports at all, as compared with only 30 per cent of the firms with 100 or more employees. But the average value of exports per employee in the small exporting firms was £14,158 compared with an overall average for all firms of £13,878. Firms with less than 50 employees accounted for 11 per cent of total exports, a share that is not inconsequential. Unfortunately, the CTT study does not give details of which products had been exported from small firms, or how the growth performance in exporting small firms compared with that of those engaged in non-traded activities.

In most other countries also, the extent of exporting tends to be lower for small firms. In the UK, for example, a survey of industry in the north-west region in 1969 showed that among potential exporters, 15 per cent of small firms (employing 200 or less) engaged in direct exporting as compared with 25 per cent for all other firms (Bolton, 1971, p. 37). Again, however, among firms that actually did export, the extent of exporting did not appear to differ greatly between large and small firms. The Bolton Report was careful to add that the indirect contribution of small firms to exporting is substantial, although difficult to estimate. This arises from the fact that exporters are significant purchasers of inputs from small firms. Japan is one of the few countries in which small firms make a very large contribution to exports. This has been made possible by the extent of subcontracting.

There are two distinct forms of subcontracting, industrial and commercial. An industrial subcontracting firm is one engaged in the manufacture and processing of industrial components to be incorporated into the product of a contracting firm. Commercial subcontracting arises where a trading company (an intermediary dealing in the distribution and flow of goods between agents) places an order with a subcontracting firm to produce a specified finished product ready for sale on the market, usually under the title of the contractor. The subcontractor is supplied with machines, materials, finance and technical assistance where required. The proportion of subcontracting firms amongst small and medium firms (employing 300 or less) in Japan has risen from 46 per cent in 1966 (Watanabe, 1970) to about 60 per cent in the mid-1970s (MITI, 1978). In 1966, a majority of subcontracting firms were contracted to a single customer firm — usually referred to as a parent company.

In regard to industrial subcontracting in Ireland, the Telesis Report (1982) noted that "very few examples exist of successful Irish companies in traded, skilled subsupply businesses, and many imports are still occurring in skilled supply businesses which should be locally sourced because of high logistics cost" (p. 121). O'Farrell and O'Loughlin (1981) found that only 11.4 per cent of the materials and components used by the largest New Industry sector, metals and engineering, were sourced domestically, though there is also evidence of an upward trend (O'Farrell, 1982). In regard to commercial subcontracting, the major effort to apply this idea in Ireland, the Irish National Trading Corporation, failed after a short time with significant losses. O'Donnell and Murray (1983) have reviewed the experience of other, smaller "development companies" in Ireland, and have suggested guidelines for the extension of this approach.

The extensive use of subcontracting in Japan, and to a lesser extent in other advanced industrialised countries, reflects the existence of a sophisticated technological and marketing infrastructure and an established tradition of inter-firm co-operation. These features tend to be lacking in Ireland. The Telesis Report (1982, p. 120) commented that:

> The skills demanded for marketing, quality control, and design and production high skilled engineering businesses are not present in most of the indigenous Irish companies interviewed during the course of our study. Irish sub-supply firms are frequently small (less than 40 people). They have limited management resources. Owner-managers frequently handle production management and scheduling, design, quality control and marketing functions. They are therefore unable to devote sufficient time to marketing or to design development, and the hiring of additional staff is frequently beyond their means.

3 *Finance*

It is widely recognised in all countries that small enterprises are hampered in their efforts to start-up or expand business because of difficulties in obtaining finance. They are too small and insufficiently well known to raise capital on the stock exchange. Conventional lenders, such as the banks, tend to cast a jaundiced eye on their loan applications because they often offer little security, may have no track record, and may be unfamiliar with the niceties of presenting a good financial case. Furthermore, in a dynamic environment, small industry will be in a constant state of flux with many new entrants and many closures, so that the risk of failure is quite correctly seen by lending agencies to be high. The difficulty cannot be adequately met by charging higher interest rates to such borrowers to cover the greater risk and handling costs of such loans (though in practice the cost of credit to small firms is generally somewhat higher than for large firms). Paradoxically, the higher the interest rate the greater the prospect that the business cannot start-up at all or will fail, since little or no operating profit may be made in the first couple of years. Small firms often have poor financial control systems, and indeed Hutchinson (1978) held that this can inhibit growth even more than lack of finance. The evidence also suggests that they are not always aware of the range of financial services available.

While the finance problem exists in varying degree for all small firms, it is particularly acute for the growing small firms, especially those that are new. The reason is summed up succinctly by Binks (1979): "The smaller the firm, the larger the proportionate increase in capital base required to respond to an increase in demand, but the lower its ability to command loan and equity finance." The expansion of the small firm in its early stages is not normally a smooth process — rather, new orders are likely to come in discrete jumps. Moreover, because of the indivisibility of capital, the response to demand may require an even larger proportionate change in the capital base. These considerations allied to the difficulty with sources of finance mentioned in the previous paragraph can present acute problems for the expanding small firm. Furthermore, if it does not expand, it will lose customers and may not be able to survive at all.

Lack of access to equity capital can be a particularly severe deterrent in inflationary conditions to new ventures which have to rely heavily on loan capital. Normally, new ventures will not reach a break-even point for a few years. The high nominal interest rates that prevail during rapid inflation have the effect of sharply accelerating the loan repayment schedule, so that the real burden of repayment is much greater in the difficult years than would be the case in non-inflationary conditions. Even though the real rate of interest may be negative, the high repayment schedule in the early years will still be a serious impediment to new enterprises if firms cannot make loan arrangements to allow for this effect.

Inflation may also exacerbate the cash flow problems of small firms in another respect: small firms often have to make more prompt payment than they can extract themselves, and this imbalance in trade credit will tend to be accentuated in conditions of high and rising inflation (Binks, 1979).

The existence of a gap in long-term finance for small firms in the UK was first noted in the Report of the MacMillan Committee on Finance and Industry (1931), which concluded that there was a shortage of long-term capital in amounts too small for public share issue. In response, a number of finance houses began specialising in small business finance and in 1959 the Radcliffe Committee reported that the MacMillan "gap" had been closed, a conclusion endorsed by the Bolton Report (1971). The Bolton Report conceded, however, that certain financial facilities available to large firms, such as access to the inter-company loan market, were not available to small ones. Furthermore, loan finance was more costly for small firms and the terms and conditions of loans could be unsuitable for new small ventures.

The Bolton Report concluded "that there was no institutional deficiency in the finance market, that while there are some differences in the bases on which small firms and large can raise money these are mostly functions of inherent cost and scale differences ..." (p. 348). Inherent cost and scale differences include the higher operational costs of handling small loans and the greater risk associated with small projects. Institutional deficiencies may arise due to excessively cautious attitudes of financiers (e.g., they may place disproportionate emphasis on asset security without due regard to actual and potential profitability) or on administrative preferences for large firms. The Bolton Report identified a certain "information gap" among small firms about sources of long-term funds. To remedy this, it advocated the establishment of small firms' advisory bureaux (which was subsequently implemented in the form of the Small Firm Information Centres). The Wilson Committee (1979), however, took the view that further measures were needed to overcome the financial disadvantages facing small firms as compared with large. The Committee recommended the introduction of a loan guarantee scheme for small businesses. Such a scheme was introduced in 1981 and is discussed further in Chapter 6.

In practice, the Bolton Report (1971) found that small firms in the UK in the 1960s were heavily reliant on internal funds on the basis of a survey of over 2,000 small firms (employing less than 200) in 1968–69, it was found that loan finance accounted for only 14 per cent of the total assets of firms. Internally generated funds were found to be an even more important source of funds for small firms than for large. Overdrafts were the main source of short-term external funding for small firms in the 1960s. This reflected their flexibility (in amount and terms), their informality and their comparative cheapness (interest being paid only on outstanding debts). A more up-to-date study of *new* small firms in Cleveland,

England (Storey, 1982) found that personal savings was the most frequently used and most important single source of finance, the second most important being the clearing banks.

It is often argued that because of the greater dependence of small firms on bank credit, and the fewer options open to them, restrictive monetary policy discriminates against small, relative to large, firms. A review of the studies on the issue conducted in ther US and Britain, however, produced no firm conclusions on the effects of monetary policy on small firms *vis-à-vis* large ones (Carson and Scott, 1973).

Trade credit is generally an important source of short-term finance for small firms, but the Bolton Report (1971) found that on the whole small firms gave more credit than they received. Small firms often find their working capital requirements raised by excessive trade credit provision due to delayed payment, even though they have less ready access than large firms to loan finance for working capital. Furthermore, small firms were found to hold a higher proportion of their current assets in the form of cash balances than public quoted companies, and their cash balances were also higher in relation to sales. This characteristic of the asset structure of small firms reflected in part the existence of economies of scale in the holding of cash balances, but also perhaps a greater conservatism of small firms in financial matters. The results of the Bolton Inquiry survey also showed that fast growing manufacturing firms were more dependent upon external borrowing and had greater recourse to bank loans than slow-growing firms.

On the basis of empirical evidence for the 1960s, Bates (1971, p. 167) summarised the financial pattern of the typical small business in the UK as follows:

> It has fairly low profits, but retains a good proportion of them and relies heavily on its own savings as a source of funds; it makes little use of long-term capital; it depends fairly heavily on borrowing from the bank and occasionally on loans from directors; it both gives and receives trade credit, which may be a vitally important source of funds; occasionally it employs hire purchase or leasing. The biggest part of its funds, therefore, comes from its own resources and from short-term and medium-term borrowing.

The subject of a gap in finance for small firms has also been debated in the US. Some writers have argued that the small business equity gap in the US has been closed (e.g., Stoll and Curley, 1970), while others have claimed the existence of a gap in medium and long-term loans (Garvin, 1968). The United States has witnessed the greatest development of venture capital institutions to meet the finance needs of small firms, especially those with growth prospects. These institutions generally take equity holdings in a portfolio of firms, which are thought to have strong growth potential. It is recognised, of course, that in the event there will be many failures and only a limited number of high-flyers; but the expectation is that the small number of high-flyers will secure an adequate overall rate of return. Following the sharp cut in the capital gains tax rate from 49 per cent to 28 per cent in 1978 and 20 per cent in 1981, there has been an explosion in the growth of venture capital investment in the US, which rose more than five-fold from 1978 to 1983, with over half of the total concentrated in computer and other electronic activities (Gonenc, 1984).

It should be emphasised, however, that venture capital in the US is not a means of finance for the generality of small firms. It is highly selective — in 1981 the 200 largest venture capitalists supported only 800 new ventures — and it is directed only at firms that are expected to reach a scale suitable for "over-the-counter" stock listing within 5–10 years, at which stage the venture capitalists realise their investment by selling their shares. Venture capital is also not very venturesome: the really risky projects are financed, initially at least, mainly from own savings or from family sources. This is not surprising given that one-third of the funds for venture capital in the US in the three years 1982–84 came from pension funds.

The evidence for the existence of a finance gap in Ireland is somewhat mixed. The various surveys of new enterprises reviewed in Murray (1981) all cited finance as the major initial problem, and in most cases the start-up finance had to come from family sources. A study by the NBST (1982) reporting the results of a separately commissioned survey relating to the finance of innovation in small firms, concluded that a finance gap does exist, especially for equity capital for start-ups where a high level of risk is involved. The study also found, however, that there was no major manifestation of demand for finance from small firms, but argued that there may be a strong latent demand which would be manifested if more suitable sources of funds could be provided. As in other countries, Irish financiers have expressed difficulties in dealing with small firms. Proposals for loan finance prepared by small firms "are often badly thought out with unrealistic assumptions of costs and sales ..." (ibid). The financiers also found that small firms lacked planning, had short-time horizons, were often inadequately capitalised and had serious working capital problems, frequently using VAT and PAYE as forms of extended credit.

O'Connor and Lyons (1982) found in a survey of 28 new small Irish firms that promoters of new ventures do not use the full variety of sources of financial support available.[21] Overall, there was a clear lack of awareness of sources and of the type of finance provided by different agencies. The Associated Banks were the main sources of financial support. Some firms expressed a reluctance to seek bank finance because they feared a breach of confidentially and also because they regarded banks as insufficiently experienced in evaluating specialised projects. O'Farrell (1985b) reported that 37 per cent of the firms in his sample of new enterprises could not have begun without the availability of the IDA capital grant, while in a further 37 per cent of cases, the grant influenced the size and scope of the business.

The most comprehensive study of the financing of Irish industry is contained in NESC (1984). The consultants' report in that publication identified the shortage of private equity investment as the key financial problem of indigenous industry, large and small. This deficiency was attributed by the consultants to low profitability in indigenous manufacturing firms and to the tax biases favouring investment in less risky assets, such as government securities and property. Venture finance was provided mainly by the state industrial grants, a form of quasi-equity which, while attractive to firms receiving them, were costly to the state. Firms were also found to be overly dependent on bank borrowing, with adverse effects on their cash flow. In evaluating loan requests, lending institutions placed the main emphasis on security in the event of default rather than on cash flow and ability to service the loan. This created a bias against new ventures, and against investment in non-tangible assets, such as product and market development. While these conditions applied in indigenous manufacturing firms generally, the consultants found the financial structure of large firms to be weaker than in small. Profitability was lower in the larger firms and debt/equity ratios much higher. Nevertheless, small firms found it harder to obtain loans from banks, the reasons "ranging from a lack of sophistication on the part of the small business to a lack of understanding on the part of the banks".

4 *Management*

Most small firms are owned and managed by the same person with little additional professional assistance. The manager will typically combine in himself or herself all functions of management as well as, in the very small firms, spending some of his time working on the factory floor. This can have undoubted advantages in keeping overhead costs down but it can give rise to many problems, because even though the manager may have practical expertise in one or two key areas he may be quite inexperienced in, and ignorant of, other aspects of business. These deficiencies are difficult to overcome since the manager is often so vital to the day-to-day functioning of the firm that he cannot take time off to

[21]The survey covered three categories of first-time entrepreneurs: those assisted under the Enterprise Development Programme and the Small Industries Programme, and those who began without any State grant. Random samples of the first two categories were drawn while the third category was derived from a number of sources. The small size of sample should be borne in mind in regard to generalisation of the findings.

attend even short-term management courses, assuming he were aware of such services and conscious of his need for them. In fact, there is evidence of a marked lack of awareness among small firms of the services available to them from different agencies, public and private, and they also find it difficult to keep abreast of technological developments.[22]

As the very small firm grows, the need usually arises for a more delegated managerial structure with professional input in areas where the owner/manager is less skilled. Many small firms fail to overcome this hurdle — for several reasons. The manager may fail to identify the need, he may lack the knowledge of how to cope with it, or he may even be unwilling to move from essentially one-man control. There is also a reluctance on the part of good quality middle management to accept employment in small firms, which often offer less security and prospects for promotion than larger corporate structures.

A survey of 166 Irish industrial firms in 1973 by Gorman *et al.* (1974) found that 75 per cent of firms employing between 25 and 99 persons ("small") were family businesses, compared with 39 per cent in firms employing 100–499 persons ("medium"), and 10 per cent in "large" firms (employing 500 or more). Furthermore, it was found that 88 per cent of the small firms were private or non-quoted companies, compared to 65 per cent of medium and 31 per cent of large firms. These findings indicate the more personal and individualistic nature of management in small firms. Gorman and Molloy (1972) observed that management in small firms was frequently characterised by the absence of rigid barriers of specialisation, a greater degree of informality in relationships, and a larger amount of sharing of information with the workforce in the firm. The absence of a managerial structure can facilitate flexibility and responsiveness to consumer needs.

These factors may also explain why industrial disputes tend to be less in small firms. Prais (1978), for instance, found in the UK that "The burden of strikes in large plants is thus heavier in all three respects: the chance of having a strike-free year is lower; the expected number of strikes per year is greater; and the number of days lost per employee is greater." Prais explained these results, however, not in terms of different patterns of individual behaviour in large plants as against small, but rather as a pure diseconomy of scale, summed up in the principle "the bigger the crowd, the greater the chance of a fight". Others have argued, however, that the relationship between size and industrial conflict is much more complex (e.g., Curran and Stanworth, 1979). Marginson (1984) argued that the effect of plant size on industrial relations is influenced by the size of enterprise to which the plant belongs. Storey (1982) pointed out that small firms are less

[22]Small high-technology firms, however, are less likely than other small firms to share the above characteristics, and tend to more nearly resemble large firms in their management sophistication.

unionised, and that strikes are positively related to the degree of unionisation. It is reasonable to expect, however, that disagreements can be resolved more often in small firms through means other than strikes, because of closer communication between management and workforce.

Small firm management is frequently found to be relatively lacking in sophistication and training. Gorman *et al.* (1974) found that the proportion of small firm managers in their sample without a formal qualification — academic, technical or professional — in 1973 was 64 per cent. The corresponding proportions in medium and large sized firms were 54 and 40 per cent, respectively. These findings, however, represent an improvement on the picture given in Tomlin (1966) for 1964, when 80 per cent of managers in small firms had no formal qualification, as against 71 and 56 per cent, respectively, in medium and large firms. Of those with formal qualifications, the majority in all size classes did not have a university degree in either 1964 or 1973. Moreover, the increase in the proportion of qualified managers between 1964 and 1973 in all size classes was overwhelmingly due to an increase in non-university professional qualifications. Gorman *et al.* (1974) stated that Irish managers were similar to British managers in the extent to which they held formal qualifications, although they were less likely to be formally qualified than managers in the US or other European countries. The authors also found that in 1973 the proportion of managers without any formal training in management was 38 per cent in small firms, 18 per cent in medium firms, and 16 per cent in large firms. The proportions had been considerably higher in 1964; 86 per cent, 80 per cent and 39 per cent, respectively (Tomlin, 1966).

The deficiencies in management technique in small firms have been illustrated by the study of Gorman, Hynes, McConnell and Moynihan (1975). They found that 46 per cent of small firms in their sample had no company plan whatsoever, and only one-quarter had a plan extending beyond a year. By contrast, 12 per cent of medium and six per cent of large firms had no plan. Similarly, the level of forecasting and planning of investment was relatively low in small firms, and tended to be of a shorter duration than in larger firms. One in three small firms had neither a system of budgetary control nor of costing. The frequency of financial reporting in trading and profit-and-loss accounts was considerably less for small firms than for large. Small firms were significantly less likely to have a separate production planning and control department than large. Poor skills in financial management in small firms were also found in the UK (Bolton Report, 1971).

O'Reilly, Jones and Coldrick (1980) highlighted the need for managers in all firms to keep pace with the flow of new knowledge and innovations in their field of activity. The problem of managerial obsolescence may affect small firms relatively more due to their lower participation in management courses, and their problem in finding time to study the relevant trade and technology journals. Even when services are made available specially to help small firms, they may not be used. The NBST (1982) study found that there was a marked lack of awareness among small firms of the services available to them from different agencies, public and private. Only about one-third of the respondent firms were aware of the services provided by the Irish Productivity Centre to small businesses in preparing proposals for expansion, while only a similar proportion were aware of the services of the IMI. Although the question to which these responses were given is somewhat ambiguous, the findings are in general accord with those published by the CII (1978) showing that for the period 1976–77, a minority of small firms were using the services provided by the different specialist organisations catering for small industry.

In sum it would appear that there are significant economies of scale in the use of managerial talent, and that the lower level of specialisation of management functions in small firms can be a serious constraining factor. Against this, the character of management in small firms may allow greater flexibility in responding to market needs as well as a higher commitment on the part of the firm's workforce. The managerial efficiency of small firms relative to large is a great extent determined by the product area in which they are engaged and the nature of the market. Small firms producing goods which require a high degree of specialisation to suit customers' needs, or which require servicing and more personalised marketing, may possess a more suitable managerial structure for these types of products.

5 *Conclusions*

We have considered the performance of the major business functions — technology, marketing, finance and management — and reviewed evidence on how well small units fare in regard to each of these functions. Small firms are often at a disadvantage, but there are sometimes offsetting advantages, depending partly on the nature of the market and the product involved. It is clear that, as Jackson (1979) emphasised, "small enterprise cannot be treated simply as a microcosm of the large".

Nevertheless, the fact that economies of scale in the discharge of the major business functions are pervasive, if not always as great as has been imagined, does raise the question as to why small manufacturing firms continue to exist in such large numbers, a question we pursue further in the next chapter.

Chapter 4

THE STRUCTURE AND GROWTH OF SMALL INDUSTRY

In this chapter, we consider a number of issues relating to the structure, efficiency and growth of small industry. But first we take up the question raised at the end of the last chapter as to why, given the prevalence of economies of scale, so many small firms continue to exist at all.

1 *Diversity of Small Firms*

In discussing the question, why small firms continue to exist, the great diversity within the small firm sector itself must first be emphasised. Some are making consumer goods, others producer goods; some are engaged in handcrafts, others in mechanical processes; and some supply many customers, while others may be supplying only a single large firm. In reducing this complexity to some sort of order, it will be helpful to categorise the reasons for the existence of small firms under three headings (i) small minimum economic scale of production; (ii) market influences; and (iii) dynamic factors.

(i) Minimum Economic Scale

Economic theory suggests that the optimum size of plant or firm is at the minimum of the long-run average cost curve, and that there will be a tendency for firms to approach that scale. It is recognised, however, that the optimum scale is highly variable from one activity to another. Moreover, it may not be a single point, since the long-run average cost curve may be flat over a wide range. In that event, firms of different scale can co-exist even in the long run once they reach the minimum economic scale.

What factors influence the minimum economic scale of production? There is large literature on this subject, which has been summarised in our previous work (Kennedy, 1971) and need not be reviewed in detail here. We confine ourselves to indicating the chief conditions which permit economic operation on a small scale.

First, low capital intensity is generally recognised as important. Some types of capital equipment are just not available in small units, and some other types, while available in smaller units, involve a high unit cost of operation. For some types of equipment the cost of production or operation is a function of surface area, whereas capacity is a function of cubic capacity, so that costs rise less than in proportion to capacity. These factors all tend to make for a high minimum economic scale in activities with a high capital intensity of production. Second, low fixed cost is conducive to small-scale operation. Where an activity involves a high fixed cost that must be undertaken regardless of the amount produced (e.g., designing an aeroplane) then the minimum scale will tend to be high. Third, activities that involve non-repetitive tasks and personalised skills favour small units. Where tasks are simple and highly repetitive, mass production methods generally become possible and make small scale production uneconomic.

These considerations suggest that the minimum economic scale will tend to be small in activities with low capital intensity, with low fixed costs, and with a high degree of personalised skill, and that small firms can be viable in such activities. Moreover, it is important to note that the scale at which unit cost is minimised is not necessarily the same for all dimensions of cost. Thus, for example, while technological considerations might point to a large minimum scale, there may be offsetting advantages in communications, labour relations or flexibility in operating on a smaller scale; and these may be more important at some stages of the production process than at others. Where different stages of the production process can be separated, this may lead to the parcelling out of some activities on subcontract to small firms to avoid the diseconomies of scale that apply to such dimensions of cost in large firms.

(ii) Market Influences

Market factors may enable firms below the minimum economic scale of production to survive indefinitely. This will be so where markets are localised by reason of high transport costs, or because a high degree of personalised service in delivery is necessary. This explains why so many service activities are carried out on a small scale. But the same considerations can apply to certain manufacturing products, particularly specialised consumer goods for which the total size of the market in any area is small. Chamberlin (1956) has shown formally that where there is product differentiation in consumer goods and entry is free, firms will be in equilibrium at less than the minimum economic scale of production.

For other goods, however, even when access to the market is not restricted, there may be substantial economies of scale in buying or selling, which will tend to make for a large size of firm, and perhaps also of establishments within that firm. In turn this factor, as well as the existence of other forms of scale economy, might give rise to oligopolistic concentration, restricting access to distribution networks. It is generally held that small firms will be at a disadvantage where advertising is an important feature of the marketing of goods.

Even though there may be substantial economies of scale in marketing, however, this does not necessarily rule out small firms from taking advantage of them while maintaining an independent identity. We have seen in the previous chapter how the prevalence of commercial subcontracting in Japan has enabled small firms to compete successfully even in international markets. It is also interesting to note that Prais (1976, p. 86) concluded that while marketing factors played a role in accounting for increased concentration of industry, it was a supporting rather than a leading role.

Moreover, market demand is rarely stable or fully predictable. Large firms may economise on their own production capacity by subcontracting to smaller firms during periods of unanticipated increases in demand or at seasonal peaks.

(iii) Dynamic Factors

The industrial structure is always in a state of flux, so that even if there is a tendency for firms below minimum economic scale to disappear, nevertheless at any given time such firms will exist. Some may be small firms that are growing into large ones. Small scale industry provides a training and testing ground for the emergence of larger enterprise: as the Bolton Report (1971, p. 29) put it, "Almost all the present large firms started off as small firms and grew, in one way or another, to their present size." Other small entrants that do not manage to grow may eventually go out of business. Again, the small firm sector at any given time will include some formerly large firms in a state of decline.

Moreover, the minimum economic scale is not fixed for all time, but can change considerably over time. One major factor is the life cycle of products, already mentioned in the previous chapter, which suggests that the minimum economic scale will be lower at the early stages of the cycle but will increase as the product matures. This would seem to suggest that new small firms are more likely to grow large if they start at an early stage of the life cycle. In mature industries, where established competitors benefit from advantages of large scale, small firms may still be able to carve out a specialised niche but they are less likely to grow large from a small beginning.

Technological change also influences the minimum economic scale. Up to recently, a common conclusion of studies of the impact of technology on size of plant or firm was that economies of scale were increasing over time (Pratten, 1971). Developments in the 'seventies, however, have raised questions whether this conclusion, if once valid, continues to be relevant. Bollard (1983), for instance, argues that small firms face relatively more favourable prospects than before, especially in sectors dependent on new technologies. Technological advances in electronics, for example, have broadened the range of commercial use of systems with the effect of generating more scaled-down and flexible modern machinery especially suited to small batch production. Developments in automation are at an early stage compared to electronics. However, the development of "computer-numerical control" (CNC) machine tools has had the effect of reducing set-up costs for short production runs, thus transferring some of the cost advantages of medium batch and mass production to small batch jobs. Although these machines are very expensive to buy, more basic CNC Models are being developed.

Prais (1976) noted that among the manufacturing trades where small firms (defined as those with less than 10 persons engaged) were gaining in Britain over the period 1958–68, many were "modern", such as scientific instruments, electronic apparatus and plastics fabrication. He concluded that "technological developments thus remain of importance in promoting small firms" (p. 15). He also drew attention to the fact that the emergence of production possibilities on a larger scale tended to give rise to a need for specialist producers operating on a small scale. His overall conclusion on the effect of technology on the scale of *enterprises* was that "In contrast to the popular view ... modern production technology offers little by way of explanation of increased concentration."

Another factor, indirectly related to technology, which may favour small-scale industry in the long run is the rise in the real price of energy. Rising transport costs may favour more decentralised production, and small firms may benefit from the fact that large firms are intensive users of energy.

As we shall show later on, the degree of flux in the small firm sector is considerable. Much of it is associated with uncertainty. In new activities the minimum economic scale of operation may not itself be known. Even where known, a new firm may not be able to secure finance to begin on that scale. Those with no track record must prove themselves on a small scale before they can command resources to expand. It is quite sensible for financiers to take this view, given that a significant proportion of new entrants fail, and that it is not possible to determine with any great assurance a priori which are likely to be the successes.

The foregoing factors illustrate the great diversity among small firms, and how unwise it would be in designing policies for small firms to assume a high degree of homogeneity. We now examine the data on the degree of difference among industries in the prevalence of small firms, and consider to what extent the kind of factors outlined above account in practice for such differences.

2 *Industrial Distribution of Small Firms*

Table 4.1 gives data for Irish manufacturing, sub-divided into ten industrial groups, on the share of small establishments (i.e., less than fifty employees) in total employment in various years. Two sets of data are used — the CSO data and the IDA data — to give a picture of changes over time for the longest period possible. It will be noted, however, that the levels in the two data sets differ somewhat, as pointed out earlier, but not to a degree that would alter the broad conclusions here.

Looking at the levels in 1980 as shown in the IDA data, it will be seen that there is considerable variation among the industry groups in the importance of small industry. The wood and furniture group had by far the highest share of employment in small industry. At the other end of the scale, drink and tobacco, textiles and chemicals were well below average. The rest of the groups were grouped within a few percentage points of the average.

These groups themselves, however, are all aggregates of a number of individual industries, which differ in the importance of small industry. We list below those groups with an exceptionally low (below 10 per cent) and high (above 35 per cent) share of employment in small industry in 1980:

Low		*High*	
	%		%
Slaughtering, meat preparation	7	Grain milling	36
Sugar	½	Bread	39
Distilling	6	Malting	52
Brewing	1	Wood	58
Tobacco	4	Furniture	67
Linen and cotton	7	Soap, detergents, candles	55
Jute	8	Leather	53
Made up textiles	10		
Fertilisers	3		
Assembly of road vehicles	8		

Source: IDA *Annual Employment Survey*. Data relate to the year 1980 and are drawn from a breakdown of manufacturing industry into 45 branches.

It was suggested earlier that activities with low capital intensity were likely to be more conducive to small firms. Using Vaughan's (1980) figures for the gross capital stock at constant (1958) prices, we examined whether this factory manifested itself in an association (negative) between the small firm share and capital intensity. The correlation coefficients between capital per worker in each industry and the share of small establishments in the industry's total employment for the years 1958, 1968 and 1973 are as follows, none being statistically significant:

	1958	*1968*	*1973*
	0.161	-0.145	0.030
n	39	39	42

Neither was there any significant relation between *changes* in capital per worker and in the small industry shares between these years. Since there is direct

Table 4.1: *Share in total employment of small establishments, Irish manufacturing groups, various years*

	CSO Data			CSO Data	
	1958	*1968*	*1975*	*1973*	*1980*
	%	%	%	%	%
Food	31.5	24.7	19.0	22.3	19.9
Drink & tobacco	17.1	14.3	9.1	13.7	10.8
Textiles	12.7	10.0	15.3	12.1	12.1
Clothing & footwear	26.7	23.0	22.2	21.8	25.1
Wood & furniture	59.8	49.7	44.5	59.1	63.3
Paper & printing	22.9	22.9	25.1	30.3	27.8
Chemicals	37.0	28.4	19.4	15.7	15.9
Minerals	24.4	18.4	15.1	22.8	22.5
Metals	19.9	13.4	18.3	23.1	25.5
Miscellaneous	36.8	27.0	25.7	26.2	25.1
Total	26.4	20.8	20.4	23.3	23.7

Sources: Irish Statistical Bulletin and IDA *Annual Employment Survey.*

evidence that small firms do in fact tend to have low capital intensity (see next section), the absence of a relationship in the Irish data at industry level presumably means either that small firms are more heavily concentrated in activities *within* industries that are generally less capital intensive, or else that they perform similar activities to large firms with a lower capital intensity.

In regard to the changes over time shown in Table 4.1, it will be seen that the share of small industry has fallen in most groups since 1958, with particularly large declines in food, drink and tobacco and chemicals. No industry group has experienced a substantial rise, but there was some increase in the most recent period, between 1973 and 1980, in clothing and footwear, wood and furniture and metals. The changes in the overall share of small industry in total manufacturing were not much due to structural change. There was a tendency for the groups with a high proportion of small industry to increase their share in overall employment, but it was slight. Taking the period 1958–75, when there was an overall fall of 6.0 percentage points, and weighting the 1958 small industry share in each group in Table 4.1 by the group's total employment share in 1975, the intersectoral change in the overall share of small industry amounted to a rise of 0.1 percentage points. For the period 1973–80, a similar exercise using a 42 industry breakdown, yielded an inter-sectoral rise of 0.3 percentage points.

It is of interest to examine whether there is any similarity in the relative degree of importance of small industry in the same sector in different countries. Because of the differences in classification, etc., between countries, it is difficult to assemble such data for a wide range of countries. In Table 4.2 we show data for Ireland in comparison with two other countries, the UK and Norway. The UK is a country with a very low overall small industry share, whereas Norway's is relatively high. To compare the relative size in the different sectors we have shown also in Table 4.2 the ratio of the sector's small industry share to the overall small industry share in each of the countries. The classification used for all three countries is the NACE code, which differs from that in Table 4.1 — hence the differences from earlier tables in the Irish figures for groups with the same title.

The data suggest that there is a high degree of concordance between the UK and Ireland in the relative importance of small industry in the same industry groups. In most cases, those groups with an above average small industry share in Ireland have an above average share in the UK, and the same holds true for those with below average shares. The chief exception is oil refining and

Table 4.2: *Shares in total manufacturing employment of small establishments, NACE groups: Ireland, UK and Norway*

	% Share			*Ratio of sector's share to country's total share*		
	Ireland (1975)	*UK (1979)*	*Norway (1980)*	*Ireland*	*UK*	*Norway*
Food, drink & tobacco	17.0	8.6	43.2	0.83	0.64	1.52
Textiles	12.9	10.5	33.4	0.63	0.78	1.18
Clothing & footwear	23.9	20.8	42.5	1.17	1.54	1.50
Leather & leather products	23.0	40.0	65.6	1.13	2.96	2.31
Wood & furniture	52.0	39.3	52.1	2.55	2.91	1.83
Paper, printing & publishing	25.1	20.9	23.2	1.23	1.55	0.82
Chemicals & allied	19.7	6.3	11.2	0.97	0.47	0.39
Mineral products	17.2	13.7	32.6	0.84	1.01	1.15
Metal processing	6.8	6.8	3.7	0.33	0.50	0.13
Mech. engineering & office machinery	27.1	16.5	21.9	1.33	1.22	0.77
Electrical engineering	13.5	14.0	6.3	0.66	0.47	0.49
Instrument engineering	19.7	14.9	31.7	0.97	1.10	1.12
Vehicles & ships	7.7	4.1	16.5	0.38	0.58	0.30
Oil refining & petroleum products	27.1	5.1	16.8	1.33	0.38	0.59
Metal manufacturers	31.2	27.9	41.2	1.53	2.07	1.45
Other manufacturing	27.6	15.6	36.1	1.35	1.16	1.27
Total	20.4	13.5	28.4	1.00	1.00	1.00

Sources: Ireland: *Irish Statistical Bulletin*, September 1981, Supplement.
UK: "Report on the Census of Production, 1979, Summary Tables", *Business Monitor*, P.A. 1002.
Norway: *Industrial Statistics Data 1980*. Data relate to establishments with 5 or more persons.

petroleum products which has a high proportion of employment in small establishments in Ireland, but an exceptionally low proportion in the UK; there are in all, however, only seven establishments in Ireland in this industrial category so that the comparison may not be of any significance. The leather and leather products group, though above average in both countries, has a far higher small industry share in the UK than in Ireland. Comparing Norway and Ireland, however, there are somewhat greater differences. Small industry is far more important in Norway, both in absolute and relative shares, in food, drink and tobacco; textiles; clothing and footwear; leather and leather products; and mineral products. On the other hand, the absolute and relative share of small industry is higher in Ireland in chemicals; mechanical engineering; and oil refining and petroleum products. The paired correlations between the small industry shares in the 16 groups in the three countries are as follows:

	Ireland	*UK*	*Norway*
Ireland		0.720	0.525
UK			0.817
Norway			

All three correlations are statistically significant at the five per cent level, but that between Ireland and Norway is the weakest of the three.

While there has been a grea deal of qualitative discussion of the matter, there has been very little research in any country to explain quantitatively the relative importance of small industry across different maufacturing industries. True, much qantitative analysis has been devoted to differences among industries in the degree of concentration. But this research focuses primarily on the relatively large firms, and only very indirectly throws light on the degree of importance of small industry.

A recent study by White (1982), however, attempts to tackle the issue directly for US manufacturing data. White regressed the proportion of sales accounted for by small firms (i.e., those with less than $5 million in sales) in 115 manufacturing industries on a number of variables. He succeeded in accounting for close to half of the variance. Those variables which turned out to be significant with the expected sign were the degree of capital intensity (–), the recent growth of the industry (+) and the distance from market (–). The rationale of the first and third of these variables is obvious enough. The recent growth of the industry, it was argued, might be used as a proxy for the newness of the industry, and more small firms would be expected in newer industries. Alternatively, rapidly growing industries may attract more new entrants. Another variable found by White to be significant, with a positive coefficient, was the fraction of sales accounted for by consumer goods. Variables used which proved not to be significant were the ratio of value added to sales (a measure of the degree of vertical integration) and the ratio of advertising to sales.

We undertook a similar analysis to try to explain inter-industry differences in the share of small industry in total employment in each industry (S) in 1968. The explanatory variables used were as follows: gross capital stock per worker (K/L, £000); the growth of the industry's employment for 1958–68 (G); the ratio of merchandise exports to gross output in each industry (D), as a measure of distance from market; and the ratio of household consumption expenditure to gross output (C), as a measure of the fraction of sales accounted for by consumer goods. The regression equation is as follows, with t values in parentheses.

$$S = \underset{(2.58)}{0.0333} - \underset{(0.42)}{0.07914K/L} - \underset{(1.29)}{0.00917G} + \underset{(1.13)}{0.01693D} + \underset{(0.44)}{0.00041C}$$

$$\bar{R}^2 = 0.0864$$
$$n = 42$$

Though the value of the multiple correlation coefficient is significant at the five per cent level, the proportion of variance explained is trivial. Moreover, none of the explanatory variables is significant, and two of the variables — the recent growth of the industry (G) and the distance from market (D) — had the opposite sign to that found in White's analysis.

This attempt to quantify the factors responsible for variations among Irish industries in their small firm employment shares is obviously not successful. Further light on the viability of small firms may be derived, however, from examination of their relative efficiency and capacity for growth, which are discussed respectively in Sections 3 and 4 following.

3 *Productivity and Efficiency*

In Chapter 3 we considered the problems and opportunities facing small firms in the performance of each of the major business functions. What matters to the viability of a firm, however, is the overall effectiveness with which a firm conducts the mix of business functions, and this will tend ultimately to show up in measures such as levels of productivity and profitability. We now consider the evidence on the relative experience of small firms on the basis of a number of such indicators bearing on efficiency.

Labour Productivity

If productivity is measured in terms of net output per person engaged, then it will generally be found that productivity is much lower in small establishments than in large. Table 4.3 gives data for Irish manufacturing on levels of net output

Table 4.3: *Average net output per head, current prices, in different establishment size classes, Irish manufacturing, various years*

Establishment size class	*1958*	*1968*	*1975*	*% change 1958–75*
	£	£	£	%
Small	609	1,363	4,399	722.3
Small-medium	642	1,575	5,047	786.1
Medium-large	760	1,748	5,914	778.2
Large	966	1,939	5,792	599.6
All Manufacturing	728	1,664	5,290	726.6

Source: Irish Statistical Bulletin.

per head at various dates in different size categories of establishments. It will be seen that in all years the average net output per head increases with increasing establishment size, and that in small establishments it was nearly one-fifth less than the mean for all manufacturing establishments. Table 4.4 shows the distribution of establishments by net output per head within each establishment size class in 1975. Only 17 per cent of small establishments had a net output per head greater than £5,000 in that year (the mean for all manufacturing was £5,290), as compared with 46 per cent of large establishments. Almost 45 per cent of small establishments had a net output per head below £2,500, or less than half the overall mean, as against less than eight per cent of large establishments.

Table 4.4: *Distribution of establishments in different size classes by net output per head, Irish manufacturing, 1975*

	Net output per head (£000 current prices)					*Absolute No. of*
Establishment	<2.5	2.5–5.0	5.0–7.0	7.0+	*Total*	*Estabs.*
	%	%	%	%	%	
Small	44.9	38.0	8.9	8.2	100	2,464
Small-medium	23.1	48.0	13.8	15.1	100	661
Medium-large	9.1	50.3	21.7	18.9	100	143
Large	7.7	46.2	23.1	23.1	100	52
All Manufacturing	38.4	40.6	10.7	10.3	100	3,320

Source: Irish Statistical Bulletin, Supplement, September 1981

Interestingly, however, as may also be seen from Table 4.3, the *growth* of average net output per head in the small size class from 1958–75 was almost identical to the growth in the overall mean, and not substantially below the growth in any other size class — in fact it was rather higher than in the large size group. The figures, of course, are in current prices and there are no separate price deflators available for the different size groups. In the absence of contrary indications, it would not be unreasonable to assume that the price changes have been much the same on average for all size classes of establishments; in which case it would follow that the growth in volume of net output per person in the small size classes kept pace with the overall rate for manufacturing. The evidence, therefore, while not conclusive, does not indicate that the productivity gap between small and larger establishments in Irish manufacturing has widened over the last twenty years.

Foley (1977) undertook a more detailed examination of the size-productivity relationship in Irish manufacturing at three different levels of aggregation and for three data samples. While the analysis for total manufacturing suggested a strong positive relationship between the two variables, the results were not so conclusive for lower levels of aggregation. The results divided roughly evenly into two main groups of industries: those where the size/productivity relationship was positive, and those where there was no relationship. Only in a small minority of cases was there any suggestion of a negative relationship between size and productivity. Foley also looked at another size/productivity relationship — the relation between productivity in the smallest and largest sized establishment classes. In almost all cases, this relationship was positive, though the highest productivity did not always occur in the largest size class.

The productivity gap between small and larger establishments is not particular to Ireland; it shows up in the available data for manufacturing in all countries (see, for example, Bolton Report, 1971). Neither does it necessarily imply that small establishments are less efficient. Net output includes, in addition to value added, certain expenses of production, such as advertising, accounting, printing, etc., which are likely to be relatively higher in large firms than in small. But even the differences between the different size classes of firms in terms of true value added per worker do not necessarily mean that small firms are economically less efficient. Economic efficiency must take account not only of the physical relationships between output and inputs, but also the relative costs of inputs, which, as we shall now see, tend to vary with size of firm.

Wages

Wages are generally found to be lower in small firms. Table 4.5 shows this for data on wages and salaries per head for Irish manufacturing in 1979, and the findings for other years are similar. It will be seen that average labour earnings

Table 4.5: *Average salaries and wages in different establishment size classes, Irish manufacturing, 1979*

Establishment size class	(1) *Salaries and wages per head*	(2) *Remainder of net output per head*	(3) *Total net output per head*	(4) *Net output per £ of salaries and wages (3):(1)*
	£	£	£	£
Small	3,567	4,784	8,351	2.34
Small-medium	4,169	6,621	10,790	2.59
Medium-large*	4,756	7,815	12,570	2.64
Large*	5,586	5,049	10,635	1.90
Total manufacturing	4,455	6,081	10,536	2.37

*The data relate to industry, including mining, electricity, gas and water as well as manufacturing. Most of the non-manufacturing data, however, were non-attributable by size of establishment, and are omitted above; so that the data are very close in coverage to manufacturing.
Source: CSO (1984).

increase with size class of establishment. Thus the degree to which small establishments fall short of the overall mean in terms of remainder of net output (i.e., net output minus wages and salaries) per head is much less in *absolute* terms than for net output per head. Indeed if productivity were measured not as net output per head but as net output per £ of wages and salaries (Column 4 of Table 4.5), then the average level in small firms, 2.34, is only slightly below the overall mean, 2.37, and is greater than the figure of 1.90 for the largest size class.

Similar findings on lower wage earnings in small firms are reported for other countries in, for instance, Shinohara (1968a), the Bolton Report (1971), Kleinman (1971) and MITI (1978). A variety of explanations has been offered in the literature. One possible explanation lies in compositional effects; for example, that small industries are more heavily concentrated in low-paid activities. The Bolton Report (1971), however, concluded that the lower average earnings in small firms are for the most part due to lower wage rates for similar activities. In Japan, however, where the wage gap between small and large firms is wider than in Western countries, Shinohara (1968a) noted that a significant factor was the seniority system operating in large enterprises. Employees of large firms in Japan tended to remain with the same enterprise where wages were geared to length of service, while in small firms this system was less characteristic and there was a high degree of mobility of employees.

Another possibility is that while the activities may be similar, small firms use lower quality or less skilled workers to perform them. It is difficult to find hard evidence on this, but to the extent that it is true it suggests that small firms may have a role in enabling people to be hired who might not get a job at all in larger enterprises. Small firms are also much less highly unionised, so that there may be greater scope for "exploitation" of labour. As against this, however, it has been argued that the job satisfaction of working in a smaller enterprise (e.g., more personalised work relationships, greater variety of tasks) are such as to compensate for lower wages and poorer facilities (Storey, 1982).

Capital Intensity and Capital Productivity

Finally, it has been argued that the lower wages in small firms are related to lower capital intensity, as measured by the capital/labour ratio. This reasoning reverses the usual text-book approach where factor prices are given to the firm, which then maximises profit by choosing the best combination of labour and capital: those with low capital intensity would have low productivity but this would show up in low profits per person engaged rather than in low wage rates, which are taken as given. Once the standard assumptions are dropped, however, and factor prices allowed to vary among firms, then a chain of causality can be established from capital intensity to wages. For instance, trade unions can exert strong pressure on highly capital intensive firms to pay above the general rate, because the cost of closure relative to the cost of extra wages will be much higher and because profits per worker would tend to be greater.

Even if capital intensity is not a factor influencing wage differentials, it is generally accepted that it would influence productivity differentials. In fact the evidence suggests that small firms usually have a much lower capital intensity than large. The Bolton Report (1971) after examining a number of different measures, concluded that capital intensity rises with the size of firm, and that capital expenditure per unit of labour input in small establishments was from 22 to 44 per cent less than in large. One factor underlying the difference is a greater tendency for small firms to use second-hand machinery. Shinohara (1968b) concluded that in Japan in the 1950s, second-hand machinery accounted for up to 80 per cent of the capacity of small firms.

If the gap between small and large firms in capital intensity were the same as the gap in labour productivity, then it would necessarily follow that capital productivity would be the same in both. In general, however, it has been found that the gap in capital intensity is wider, so that capital productivity tends to be greater in small firms — see, for example, Economic Planning Board of Japan (1957), Bolton Report (1971), Ahmed (1976) and World Bank (1978). This result accords with what would be expected from standard marginal productivity theory.

Profitability

The evidence generally suggests that while labour productivity is lower in small firms, the effect on costs is offset by lower wages, lower capital intensity and higher capital productivity. Each of these indicators is only a component of cost, however, and none of them on its own tells us whether efficiently is less or greater overall in small firms. One indicator that might be used to assess the net outcome of the other components is profitability as measured by the return on capital. While in one sense this is no more than another partial indicator, it is, nevertheless, a particularly crucial one because of its influence on the inducement to invest, and therefore on whether an enterprise will exist at all or not.

There have been many studies of the relationship between size and profitability, and the results are by no means clear-cut. This is hardly surprising given the complexity of the concept of return on capital and the different possible ways in which it can be measured. Using financial data, rates of return can be calculated on equity (assets less liabilities), net assets (equity plus long-term loans) or total assets (net assets plus current liabilities). Different accounting procedures, however, may artificially affect the comparison of rates of return in different firms. Large firms tend to have more sophisticated accounting procedures and are more likely to value fixed assets at replacement cost, whereas small firms may continue to value them at purchase cost. Economic estimates of the value of capital stock are more likely to be on a consistent basis, but these estimates are generally done on an aggregate basis or by industry, and are rarely available for different firm sizes. Satisfactory measures of the numerator, the net return to capital, are also difficult to derive. Apart from the fact that different methods are used to estimate depreciation, there is also the problem in small firms of separating the return to capital from the return to the owner-manager for his labour. For these reasons the available measures of profit are often almost meaningless in the case of small firms.

Several studies have produced evidence suggesting that profitability increases with firm size — Anderson (1967), Hall and Weiss (1967), Osborn (1970) and Bate (1971). On the other hand, Samuels and Smyth (1968) and Tamari (1970) concluded from the evidence they examined that profitability and firm size were inversely related. Singh and Whittington (1968) examined the relation between size and profitability on an industry by industry basis to avoid the possible aggregation biases that might result from a variety of sources (e.g., differing degrees of scale economies and monopoly power, varying relative prosperity of industries in different periods). They found that while profitability was generally lower the larger the firm size, the difference was not significant at the five per cent level. A large sample inquiry undertaken by the British Inland Revenue relating to companies exempted from submitting annual accounts to the Registrar of Companies — and generally relatively small — found that, while the rate of return increased with size within this sample, the average rate for all sizes within the sample was higher than in large quoted companies (Board of Trade, 1967). The Bolton Report (1971) found that the average small firm in their survey had a higher profit rate than the average quoted company. In the light of this and other evidence, they concluded that, despite the qualifications attaching to such comparisons, it was probably correct to conclude that small firms earned a higher return on capital, but that the difference was not very great.

While the evidence is far from unambiguous, it can at least be safely concluded that there is no definitive evidence that small firms are generally less profitable than large. A more clear-cut conclusion common to all studies on the size/profitability relationship is that the variability of profit is inversely related to size. A major reason underlying the greater variability of profit rates among small firms is that they depend on a narrow range of production, and are less able to offset losses in one line against profits in another. Also large firms may be better able to exert monopoly power to control the market so as to stabilise profit rates over time. The evidence of a possibly higher average rate of return in small firms, combined with greater variability in returns, would be consistent with a higher degree of risk in small firms.

Efficiency

While the foregoing indicators are of interest in their own right, and give some basis for assessing the relative efficiency of large and small enterprise, none of them can be regarded as conclusive measures of efficiency in the use of economic resources. Indeed the Bolton Report (1971, p. 47) concluded that there is "no way in which the relative efficiency with which firms use the resources they employ can be compared". Notwithstanding this stricture, however, a number of attempts have been made to derive indices of the "social efficiency" of establishments in using resources.[23] The inputs of labour and capital are weighted, not by market prices, but by weights reflecting the opportunity cost of the factors. The value of output is then related to this input measure to derive an index of social efficiency, which is essentially a measure of the productivity of resources used when these are valued at their estimated social opportunity costs (i.e., free from the factor price distortions that affect ruling market prices).

An application of this technique has been made by Ahmed (1976) to the Irish plastics industry, classifying firms into three size groups: small (1 to 19 employees), medium (20–49) and large (50+). He found that efficiency was highest in the medium-size class and lowest in the small size group. The differences between the means of any two size groups, however, was not statistically significant. While the evidence relates to only one industry, it does not support the view that large establishments are in general more efficient socially in their use of resources. Prais (1976, p. 168) in a wide-ranging study of the factors accounting for the rise of giant firms in the UK concluded that "There are general

[23]For references, see Ahmed (1976) and Todd (1977).

pressures, unconnected with efficiency, leading to the emergence of ever-larger firms," and that the benefits involved were often private rather than social. It should be added, however, that where large-scale is necessary to meet foreign competition, the benefits of large size may still be "social" for the country of location, even if not for the world economy.

Efficiency in the use of resources as measured in this way, however, is essentially a static concept. It does not therefore comprise all the aspects of the relative effectiveness of different sizes of firms, such as the ability to innovate. Another important dimension is the capacity of different sizes of firms to grow, to which we now turn.

4 *Growth of Small Firms*

There have been many attempts to establish generalisations about the growth prospects of small firms or establishments in relation to large, and a considerable literature has developed on the subject. Before going on to consider the evidence in regard to Irish small industry, it may be of interest to discuss some of the key issues that have been debated in that literature.

Much of the literature takes its starting point from Gibrat (1931) who postulated a growth process that has come to be known as the Law of Proportionate Effect. Put simply, this law states that in any given period, while growth rates vary considerably among firms, all firms regardless of size have an equal chance of a given proportionate change over the period. According to this law, the mean and variance of growth rates of firms in each size class should be the same for all size classes. This hypothesis is convenient in theorising about the size distribution of firms, and over a sufficiently long-time period suffices to explain the emergence of a high degree of concentration of activity in the largest size class. Its validity, of course, depends on empirical verification. Other research workers have turned their attention to this question and in the process have sought to develop and refine the hypothesis in various ways.

Hart and Prais (1956) concluded from examination of UK data in various sub-periods over the period 1896–1950, that abstracting from births and deaths, the distribution of growth rates was similar for all sizes of firms, thus supporting the hypothesis that small and large firms have an equal chance of a given proportionate change in size. As regards deaths (i.e., firms that ceased to exist), they found that the death rate tends to fall as firm size increases. New entrants (births) during each sub-period were found to be below average size at the end of the sub-period. These data, however, related to *enterprises* quoted on the London Stock Exchange; they included non-manufacturing enterprises; size was quantified according to market valuation; births were defined as new company quotations; while deaths referred to removal from the stock exchange lists. Thus, mergers and take-overs would lead to the "death" of an enterprise, while "births" would often have been in operation prior to the initial date of quotation. For these reasons, the results might not correspond with those based on different data coverage and definitions.

Simon and Bonini (1958) claimed support also for the law of proportionate effect using data on the 500 largest manufacturing enterprises (in terms of sales) in the US over the period 1954-56. They implied that the results would be similar for plants and firms, and whether size was measured by sales, assets, employees, value added or profits.

The results have been challenged by other scholars. Hymer and Pashigian (1962) argued that if large firms were simply aggregates of independent small firms, the standard deviation of the growth rates would be expected to decrease with increases in firm size by a factor of $1/\sqrt{n}$, where n is the measure of firm size. Using data on the growth rates of the 1,000 largest firms in the US in the period 1946-55, they found that while there was little difference in the mean growth rates for different sizes of firm, the standard deviation of the growth rates fell with increasing size, but by a factor of less than $1/\sqrt{n}$. Simon (1964) conceded that the assumption of equal variance was not in accord with the generally available facts, but argued that the data were compatible with a weaker form of Gibrat's law. Boswell (1972) also found that although there was no apparent association between employment growth and the size of firms studied by him, the smaller the size group the greater the dispersion of the average growth rates. Other studies, such as Mansfield (1962) have found that smaller firms tend to have higher death rates, while those that survive tend to have higher as well as more variable growth rates than larger firms. Even earlier, Meyer and Kuh (1957) had found a negative relationship between size of firm and both the mean rate of growth and its variance.

The most striking recent findings in that regard are those of Birch (1979) in relation to US manufacturing data, based on a data-file of individual firms at different dates. Between 1969 and 1976 when total US private manufacturing employment fell by 151,000, employment in establishments of 1-20 employees rose by 543,000, and almost two-thirds of this increase was in independent single-establishment firms. The only size classes with net gains in employment were those with up to 50 persons engaged, and employment in new firms less than five years old amounted to a high proportion of the total net increase in employment. Examining the details of manufacturing establishments which existed at the beginning, Birch's data show that the average growth rate of small establishments (i.e., either those with 20 or less employees or those with 50 or less employees) was greater than for large establishments. Variability was also greater among small establishments. For example, in establishments with 20 or less employees in 1969, 23.2 per cent had increases of 25 per cent or more in employment in the period 1969-76, while 57.8 per cent had declines of 25 per cent or more; the corresponding figures for firms employing over 500 were 13.6 per cent and 41.9 per cent, respectively. In the case of declining firms, the odds of going out of business altogether were much higher in small firms than in large: 77 per cent of declining firms with less than 20 workers went out of business altogether, as against 29 per cent for firms with more than 500 workers.

There has been much questioning of Birch's findings, particularly in regard to the extent of the contribution of small firms to overall employment growth in the US. The NESC (1983, p. 21) report has drawn attention to some difficulties in establishing the precise basis of some of Birch's results. Moreover, there seems to be an outright conflict of evidence between Birch's findings that independent single-establishment firms with less than 20 persons engaged accounted for 52 per cent of the total private sector employment increase in the period 1969–76, and the finding of Armington and Odle (1982) that small firms with less than 100 employees accounted for 39 per cent in the period 1978–80. The difference may, of course, be merely a reflection of the different time-periods involved; but if so, it constitutes a warning about generalising from a particular period. Though the difference between the two studies is substantial, it is a matter of degree, however, since Armington and Odle confirm Birch's findings that, in both the economy as a whole and in manufacturing, employment grew faster in small firms than in large. The same is true of the critique by Fothergill and Gudgin (1979), whose evidence relating to the UK suggests that small firms are a better bet for employment growth than large firms, but who also emphasise that a satisfactory overall employment growth is unlikely without a positive net contribution from the larger firms. Likewise Storey (1982), in a careful review of the issues involved, emphasises that, at a time when a net increase in employment in small firms was accompanied by a net decrease in employment in large firms, data on the *share* of small firms in the overall net employment increase tended to exaggerate the contribution of the small firm sector to employment generation.

A number of writers have turned their attention to the question of what accounts for the observed growth patterns of firms within and between different industries. Simon and Bonini (1958) attributed the growth experience of individual firms to such factors as profitability, investment, and mergers; and these variables in turn were seen as dependent on a variety of forces, such as the efficiency of the firm itself and the growth of the industry in which it was located. They postulated that constant returns to scale prevail once a minimum economic scale is reached. These factors generate a probability distribution for changes in the size of firms of a given size class. Their basic assumption then was that the probability distribution was the same for all sizes of firms above the minimum economic scale.

Given the evidence conflicting with the strong version of the law of proportionate effect, however, Ijiri and Simon (1964) described a stochastic process based on a variant of Gibrat's law that could account for the observed phenomena. The modification introduced was to allow for the possibility that a firm's growth in one period was influenced by its growth in the preceding period. The work was extended in Ijiri and Simon (1967) and tested with data for large US firms over the period 1954–62. It was found that a firm which grew rapidly in the first half of this period could expect a greater than average growth in the second half. The excess of its growth above the average, however, would be greatly reduced: 'rapidly growing firms "regress" relatively rapidly to the average growth rate of the economy' (p. 355). Ijiri and Simon (1974) concluded in relation to US firms over the period 1968–69 that "mergers and acquisitions in this period increased the degree of concentration", with smaller firms having a higher probability of being absorbed by mergers and takeovers. Samuels (1965) and Utton (1972) have concluded that in the UK in the post-war period mergers and acquisitions have been important in enabling large firms to expand more rapidly than small.

The explanatory framework developed by Simon and his co-authors essentially seeks to explain concentration patterns and differences in growth rates with a minimum of basic asusmptions. Hymer and Pashigian (1962 and 1964), however, argue that the greater variability in small than in large firms' growth points either to the presence of continual economies of scale or that large firms can establish a degree of monopolistic control — factors that are not considered necessary by Simon to explain the observed phenomena. Mansfield (1962) stressed the importance of innovation, finding that, on average, successful innovators grew about twice as rapidly as other firms in the same industry.

Birch (1979) found that the odds of contracting and expanding both fell with age of firm, indicating greater stability in older firms: this held even when size was controlled for. Boswell (1972) also found that the age of firms, as well as the age of management, were both inversely related to the growth of the firm, indicating that young firms, or firms with young managers, tended to have above average growth. Birch (1979) further emphasised the importance in determining growth rates of (i) government incentives to small firms and to industry generally, (ii) industrial relations, (iii) location, and (iv) the product and industry in which the firm is involved. The impact of these factors was not quantitatively assessed, however.

Turning to the Irish data, which it must be emphasised relate only to establishments and not enterprises, it was shown in Chapter 2 that employment in small establishments in Irish manufacturing rose by 13.2 per cent from 1973–80 as against a rise of 11.3 per cent overall. This relative increase in small industry does not, however, tell us whether small establishments were growing more rapidly than large, since the establishments that appear in one size group in any year are not necessarily those that appear in the same size group at a later year.

It is possible, however, to elucidate the issue by considering the fate of the establishments which existed in 1973. Of the establishments with less than 50 engaged in 1973, 47.1 per cent had maintained or expanded employment in 1980, as against 38.6 per cent of all other establishments. Of the firms which declined, almost half of the small establishments had gone out of business altogether as against a quarter of the larger firms. The corresponding figures for employment changes are given in Table 4.6. These show that small establishments fared better overall than others: there was scarcely any net change in employment from 1973–80 in the small establishments existing in 1973 as against a net decline of 12 per cent in all sizes of establishment.[24] The gross components that make up the net changes also illustrate the greater volatility of small industry, where the rate of gross gains and gross losses were both higher than in total manufacturing. Furthermore, the data also indicate that closures form a much higher proportion of job losses in small firms than in large.

A more precise test of Gibrat's law is given in Table 4.7 which shows the unweighted mean and standard deviation of the percentage changes in employment 1973–80 in all establishments in different size classes. The data again relate only to establishments in existence in 1973 and are classified according to size in that year. Two separate measures are given, one including all such establishments and the other only those that were still in existence in 1980. Whichever measure is used, the data show a clear tendency for mean growth to decline with size. Moreover, the standard deviation tends to fall with size, indicating the greater variability of growth in smaller establishments. In Irish manufacturing, therefore, small establishments while liable to greater variability in their growth experience than larger establishments, are more likely to achieve a high growth rate. These results do not support Gibrat's law, but conform with the patterns identified by Birch (1979) in the US data. The reasons for this conformity may differ, however, since in Ireland the reduction of protection had a more adverse effect on large firms than on small.

Finally, it is interesting to look at the overall contribution to employment change resulting from the pattern of change in small industry in this period. The data are given in Table 4.8. For this purpose we include the total of 51,790 jobs in 1980 in establishments that entered after 1973, and we classify these establishments on two different bases: A, according to their size in the year they commenced, and B, according to their size in 1980. New entrants accounted for

[24]As mentioned in Chapter 2, O'Farrell (1984) has suggested that the IDA Survey figures give an unduly favourable picture of the employment trend in small industry — because of incomplete returns for Dublin in 1973 on small establishments which had closed by 1980. If we exclude Dublin altogether from the figures, however, the relatively better performance of small establishments is not affected: outside Dublin there was in fact a rise of 4.5 per cent in employment for small establishments existing in 1973, as against a fall of 7.6 per cent in all sizes of establishment.

Table 4.6: *Employment changes 1973–80 in establishments existing in Irish manufacturing in 1973*

Establishment size in 1973	*1973 total employment*	*1980 total employment in same establishments*	*Gross changes*[a] *Gains*	*Closures*	*Contractions*	*Total losses*	*Net change*
Small	50,745	50,682	16,695	11,316	5,442	16,758	-63
Small-Medium	66,244	59,934	12,065	10,544	7,831	18,375	-6,310
Medium-Large	51,113	39,688	5,517	8,604	8,338	16,942	-11,425
Large	49,657	40,424	2,815	2,576	9,472	12,048	-9,233
Total	217,759	190,728	37,092	33,040	31,083	64,123	-27,031
			Change as a % of 1973 Total Employment in Size Group				
Small			32.9	-22.3	-10.7	-33.0	-0.1
Small-Medium			23.8	-20.8	-15.4	-36.2	-12.4
Medium-Large			10.8	-16.8	-16.3	-33.1	-22.4
Large			5.7	-5.2	-19.1	-24.3	-18.6
Total			17.0	-15.2	-14.3	-29.4	-12.4

[a] "Gains"/"Contractions" are measured as the sum of the employment changes between 1973 and 1980 in establishments which existed in both years and where the 1980 level was greater/less than that of 1973. "Closures" refer to job losses in establishments which were in employment in 1973 but had zero employment in 1980. Relocations appear above as closures (and as new entrants in Table 4.8 following) — thus arguably inflating somewhat the figures for closures (and for new entrants), depending on the reasons for relocation.

Source: IDA *Annual Employment Survey.*

Table 4.7: *Mean and standard deviation of the percentage changes in employment 1973–80 in all establishments existing in Irish manufacturing 1973*

	All establishments existing in 1973[a]		*Surviving establishments*[a,b]	
Establishment size in 1973	*Mean*	*Standard deviation*	*Mean*	*Standard deviation*
	%	%	%	%
Small	6.8	168.9	43.3	181.7
Small-Medium	-7.9	102.4	10.3	102.7
Medium-Large	-20.3	56.6	-4.6	48.3
Large	-18.8	37.7	-13.0	31.8
Total manufacturing	3.4	157.4	35.4	167.7

Notes:
[a]The means here are unweighted averages of the percentage changes in each establishment, and for that reason are not comparable with the figures in Table 4.6.
[b]Surviving establishments are those existing in 1973 which were still in business in 1980.
Source: IDA *Annual Employment Survey.*

nearly three-fifths of the total gross gains. Of these the majority — or nearly 60 per cent — began with less than 50 workers in their first year of operation; and even if we use the alternative basis of classifying new entrants, small establishments still emerge as providing the largest fraction for any of our four size groups (35 per cent). Small establishments accounted for 45 per cent of the gross gains in establishments which were already in being in 1973. Turning to losses, small establishments accounted for one-third of jobs lost through closures, the highest share of any of the four classes given in the table, but they accounted for the lowest share of employment decreases in those declining establishments that still survived.

Overall, small establishments accounted for a higher share in both total gross gains and total gross losses in the period 1973–80 than their initial employment share in 1973. Their share in gross employment gains, however, considerably exceeded their share in gross losses, so that they represented the biggest contributor to the net increase in employment. Large and medium-large establishments contributed poorly to increases in employment, and the job-generation process depended heavily on new establishments, many of which at least began small. In interpreting these figures, however, it must be borne in mind that most new firms will tend to start small. The possibility must also be considered that some of the

Table 4.8: *Proportion of employment changes 1973–80 in Irish manufacturing accounted for by establishments in different size classes*

Establishment size[a]	*Gains*					*Losses*				
	New entrants			*Total Gains*				*Total*	*Net change*	
	A	*B*	*Increases*	*A*	*B*	*Closures*	*Decreases*	*losses*	*A*	*B*
	%	%	%	%	%	%	%	%	%	%
Small	59.4	35.1	45.0	53.4	39.2	34.2	17.5	26.1	123.9	73.1
Small-Medium	30.7	31.3	32.5	31.4	31.8	31.9	25.2	28.7	38.7	40.1
Medium-Large	4.9	20.3	14.9	9.1	18.0	26.0	26.8	26.4	-35.9	-3.6
Large	5.1	13.2	7.6	6.1	10.9	7.8	30.5	18.8	-26.7	-9.6
Total	100.0	100.0	100.0	100.0	100.0	100.0	100.0	100.0	100.0	100.0
Absolute Nos.	51,790		37,092	88,882		33,040	31,083	64,123	24,759	

[a]Establishments existing in 1973 are classified by size in that year. Establishments entering later are classified on two bases. Under A, new entrants are classified according to their size in the year they entered, and under B they are classified by size at the end of the period, 1980. These two different classifications of new entrants account for the two different classifications of total gains and of net change.

Source: IDA *Annual Employment Survey*.

establishments were branches of larger firms, and it might be thought that very different results would arise if the data were given on a firm or enterprise basis. While it is not possible to establish satisfactorily how many of the small establishments were independent enterprises, a check through the names of establishments in the IDA *Annual Employment Survey* indicated that less than about 10 per cent of employment in small industrial establishments was located in the obvious branch establishments of larger firms. More important, however, the employment performance in these branches of larger firms was not superior to that of the independent establishments.

The period 1973–80 may not be typical of all phases of industrial development in Ireland, since this was a time when the older and larger firms were suffering intense competition as a result of EEC entry and the emergence of the newly industrialising countries. There were also structural factors affecting particular industries. In the face of these factors, there can be little doubt but that the performance of industrial employment would have been much worse without the contribution of the small firms. Nevertheless, as Fothergill and Gudgin (1979) stressed, since the bulk of employment is located in larger establishments, a satisfactory overall employment growth is unlikely to be achieved if, on balance, the existing large establishments are not also making a positive net contribution.

5 *Conclusions*

We have considered the factors that might account for the continued existence of small-scale industry despite the operation of powerful forces towards concentration. As might be expected some activities are more suited than others to small-scale operation — a point borne out by the fact that the relative importance of small industry in different industrial branches is correlated across countries. In general the evidence does not suggest that small-scale businesses are less efficient or have poorer growth prospects than large. A vibrant small firms sector can help to maintain a competitive environment, exercising some restraint on the monopolistic tendencies of larger firms, and providing a spur to innovation. In the latter connection, it is noteworthy that the surge of innovation in small firms in the US in the second half of the 1970s arose partly because established firms had cut back on R & D activity in the face of the economic crisis.

A word of caution should be entered, however, about the implications of such findings. It would be as foolish to claim that small industry is the answer to Ireland's problems as it would be to ignore its potential contribution. The findings are necessarily drawn from an environment in which small and large firms exist. In some cases the development of small firms is helped by the presence of large firms, which either create a market where small firms can carve out a niche or buy components from small firms. In such cases small-scale firms might not even exist without the prior presence of the large. In other cases, where small firms compete with large firms abroad, the latter may have competitive advantages either because of genuinely greater efficiency or because of the exercise of market power. In such circumstances, sole dependence on small firms would be unlikely to produce satisfactory industrial development unless the disadvantages of size could be overcome by combination with other small firms.

Chapter 5

SMALL INDUSTRY AND ENTREPRENEURSHIP

Protagonists of small industry frequently see it as an avenue by which new entrepreneurs can emerge: indeed some would go so far as to hold that it is a necessary training ground for the emergence of that function. Sceptics, however, question not only the significance of small industry for entrepreneurial development but even the importance of the entrepreneurial role itself. In this chapter, we consider what, if any, importance can be attached to the entrepreneurial function and the role of small industry in developing it.

Notwithstanding the fact that some eminent economists have devoted considerable attention to the subject, it remains true that the entrepreneur does not figure much in mainstream economics. There are many reasons for this. Mainstream economics is concerned with the functioning of developed economies, in which it is not too outrageous to assume that the market will respond readily to economic incentives. The typical production function makes no explicit reference to an entrepreneurial input — by implication treating the managerial function as a specialised variety of skilled labour, and the risk factor as an element in the return on capital. Indeed as Leff (1979, p. 60) pointed out "In an ideal market system, without uncertainty, factor market imperfections, and externalities, entrepreneurship would not be necessary." However valid this approach may be for developed economies — and even in such economies it is open to question (see, for instance, Gilder, 1981) — it is clear that the assumptions do not hold good for underdeveloped countries, or for the developed countries at an earlier phase. Not surprisingly, therefore, the question of entrepreneurship has figured much more prominently in the literature on development economics and economic history.

The neglect of entrepreneurship in mainstream economics has not been without its critics within the profession, however, and Baumol (1968, p. 66) remarked that "We are led to suspect that by ignoring the entrepreneur we are prevented from accounting fully for a very substantial proportion of our historic growth." For a country like Ireland, which is still in the process of industrialising, it would be unwise to assume a priori that the entrepreneurial role in manufacturing can be treated as an essentially reflex response to the presence of opportunities for profit.

1 *The Role of the Entrepreneur*

But what is the role of the entrepreneur? The concept has had a chequered history, leading one critic, Chandler (1962) to suggest that the difficulties of interpretation might best be overcome by dropping the term altogether! The concept was in fact first coined by an Irish-born economist, Richard Cantillon, in the eighteenth century to refer to the individual bearing non-insurable risks. There are, however, many individuals and groups who bear non-insurable risks (e.g., shareholders, speculators), but who are otherwise dormant in relation to the productive process. The notion of the entrepreneur as the one who organises and co-ordinates the other factors of production derives from the French economist, Say. For Schumpeter (1934, 1950), the entrepreneur is pre-eminently the innovator, who seizes opportunities for the introduction of new products and new techniques and so disturbs the prevailing equilibrium. The modern Austrian school, on the other hand, views the entrepreneur as one who, in a state of imperfect knowledge, spots an opportunity for profit due to a prevailing disequilibrium in the market, and so helps to bring about a movement towards equilibrium. In the context of developing countries, where markets for output, labour and capital may be very imperfect, Leibenstein (1968) sees the essential role of the entrepreneur as the "gap-filler" in exploring marketing opportunities, and the "input-completer" in combining the factors of production.

Thus the main roles assigned to the entrepreneur relate to coping with risk and uncertainty, and to the organisation of production in non-routine ways. That the importance attached to the different roles has varied over time and place is partly a reflection of the differences in the economic environments over time and place. Thus, for example, the rise of the joint-stock company served to spread risks over a much wider number of people, thereby greatly diluting the significance of the individual entrepreneur in regard to the function.

Not only do the entrepreneurial roles change, but so also do the ways of performing these roles. The conception of the entrepreneur as an individual with distinctive personality traits had to undergo radical revision in the light of actual experience. In the more developed countries in the first thirty years after the Second World War, the entrepreneurial role was largely assumed by that of the technocrat in large conglomerates, and transnational corporations extended this process to less developed countries. State-sponsored enterprise appeared in many countries, developed and underdeveloped; and while performance generally has been mixed, nevertheless developments have been initiated that would not otherwise have taken place. In some countries workers' or producers' co-operatives assumed an important role in industrial development, while there are also some notable examples of trade union industrial enterprise. In the United States, however, there has been, in recent years, a great resurgence of the smaller-scale private entrepreneur. The "push" and "pull" effects of changing economic conditions is likely to affect the mix of entrepreneurial supply (Binks and Coyne, 1983). Recessions raise the proportion of "pushed" entrepreneurs, i.e., those motivated by the actuality or risk of unemployment.[25] Boom conditions raise the proportion of "pulled" entrepreneurs, i.e., those attracted by a particular product or process idea. The "pushed" are more likely to favour an activity similar to what they have left, whereas the "pulled" may be more innovative and original.

Moreover, general economic measures aimed at overcoming constraints in access to markets, technology, finance and the other factors of production have evoked a much greater indigenous entrepreneurial response in developing countries than had been anticipated thirty years ago. Leff (1979, p. 60) goes so far as to conclude that "It has become clear that economic development in most current LDCs can proceed without those countries having to wait for a psycho-cultural transformation that would increase the supply of entrepreneurs." Likewise, Storey (1982), while accepting that psychological and sociological explanations are valuable, notes that it is not wholly clear what policies would follow from these explanations, and that at best they would be effective only in the very long term.

Nevertheless, a note of caution is in order before dismissing the possibility of entrepreneurial deficiency. First, even if the supply of entrepreneurship generally responds positively to improvements in economic incentives and removal of barriers to entry, the elasticity of supply may still vary from one society to another depending on socio-cultural factors, such as attitudes to risk and failure, the degree of accumulated industrial experience etc. A second and related point is that the level of economic incentives required in some societies to attract sufficient enterprise may result in inequalities that would not be socially tolerable. Papenek (1962) in a study of Pakistan found that while there was an enormous response to enhanced profit opportunities in a country that previously had little tradition of industrial enterprise, this was achieved through heavy protection ensuring vast profits. Annual rates of return of 100 per cent on investment were common in some industries, and "almost any industrialist was guaranteed a profit". As Papenek noted, not every social system would be capable of tolerating the co-existence of such high profitability and high prices for consumers.

Third, experience in the less developed countries shows that increasing the economic incentives to attract more enterprise has not consisted simply in making the market system operate more effectively. On the contrary, there has been massive intervention in the market system by government action —

[25]A "push" factor of a different kind is present in the high degree of entrepreneurship often noted among displaced minorities and refugees.

through trade protection, state purchasing, state enterprise, and measures to favour local groups and industries. While these measures have broken down barriers facing native enterprise in access to markets, technology and finance, they have often created extensive distortions and inefficiencies (Leff, 1979). Many protectionist measures to induce greater indigenous entrepreneurial activity, whether desirable or not on balance, are simply not an option for a small, highly open, economy like Ireland seeking to industrialise in free trade conditions.

2 *Indigenous Entrepreneurship in Ireland*

Is there any evidence of a dearth of indigenous industrial entrepreneurial activity in Ireland? Unfortunately there has been no systematic research on this question. Even to pose it — and thus raise the possibility that the answer may be in the affirmative — is sometimes regarded as a reflection on the Irish personality. But that is to confuse a possible fact with its possible cause. Even if it were established that there was deficiency of indigenous manufacturing entrepreneurship, it does not follow that the cause lies in inherent personality defects. Tradition and experience, for example, play a big part in determining the aptitudes of a population, and in particular the direction in which they channel their enterprise. Furthermore it does not suffice to reject the possibility of entrepreneurial deficiency to show that the Irish are no less acquisitive than citizens of any other country, since the maximisation of monetary returns will direct people into those areas where they can best apply their aptitudes and experience. If through tradition or whatever, they lack confidence in their own ability to derive satisfactory returns from manufacturing enterprise, even though foreigners can do so in the same environment, then they will direct their energies elsewhere to greater advantage.

There are a number of factors that would tend to support the hypothesis of indigenous entrepreneurial deficiency in manufacturing. First, as Lee (1983) emphasised, insecurity was a central feature of nineteenth century Irish life, with the not surprising outcome that "performance took second place to possession as a criterion of legitimacy". There are indications that this ideology carried over into the twentieth century, and that it is alive and well in many areas of economic life — notably in the professions, the public service and a considerable part of agriculture — that have succeeded in insulating themselves from change. It is not a climate well adapted to bringing to the fore the talents needed for successful manufacturing enterprise.

Second, there has been a relatively poor use of new industry grants by established indigenous enterprises. The package of industrial incentives has been available equally to Irish and foreign firms; Irish firms are better placed to know about them; and Irish firms as a matter of course have local knowledge about the social and economic background, which can only be acquired at some expense by foreign firms. Yet according to McAleese (1977), 60 per cent of total employment arising in the IDA New Industry Programme occurred in foreign firms. Furthermore, the bulk of the domestic projects assisted under the New Industry Programme represented expansions in firms surviving from the tariff-protected era, and only a small part represented wholly new undertakings.

While these facts are not in dispute, they are open to interpretations other than entrepreneurial deficiency. For instance, they could be explained in terms of barriers to entry or expansion facing indigenous firms in access to markets, technology or finance. In turn these barriers could be explained by market imperfections reinforced by economies of scale and external economies. Even if such an explanation is accepted, however, it does not entirely dispose of the hypothesis of entrepreneurial deficiency. It is surely part of the function of good entrepreneurship to surmount such barriers. The possibility of entrepreneurial deficiency cannot be ruled out, therefore, unless it can be shown that in Irish conditions the barriers have been so considerable that they could not be removed without the intervention of a force external to the market, such as the State.

A third pointer to entrepreneurial deficiency in indigenous manufacturing, related to the foregoing, is the apparently very different rates of profitability in indigenous manufacturing as against foreign enterprise in Ireland. McAleese (1983) quotes United States Department of Commerce figures showing an average annual rate of return on US manufacturing investment in Ireland of 31.7 per cent during the four years 1977–81. While comparable figures are not available for indigenous enterprise, it can be taken as certain that the rates of return were not remotely as high as this. As with the previous point, this fact is open to different interpretations. The high profits earned by foreign enterprise may be partly due to transfer pricing and/or to licence fees, representing a return on earlier investment in R & D.

A fourth point is that, if we are to accept the judgement of McAleese (1983), there has been no shortage of capital in Ireland, but rather a want of ideas on how to use it productively. Indeed for much of its history, Ireland has had substantial external assets, and could therefore be said to be an exporter of capital as well as labour. The lack of use, or misuse, of capital does, however, call for an explanation. McAleese sees most of the misuse as being due to misallocation by the government. The State in Ireland, however, did not intervene in capital allocation for ideological reasons, but rather for the pragmatic purpose of overcoming deficiencies in the private sector — so that it can plausibly be argued that misuse by the State would not have been an issue, were it not that private enterprise, for whatever reason, was deficient in the first place.

It may be pointed out, however, that the IDA Small Industry Programme has brought forward a very large number of new small enterprises simply by changing certain elements in the economic environment facing new entrants. That indeed is true, but it is also true that very few of these have developed into really large enterprises. O'Farrell and Crouchley (1984) concluded that "Although the rate of indigenous new firm formation in Ireland has been relatively high by international standards, and very high relative to the UK, there is evidence to suggest that most of the firms established are small concerns which are very unlikely to expand into even medium-sized enterprises selling overseas." The evidence only demonstrates that entrepreneurial competence exists at a small scale; it has yet to be demonstrated that it exists for the more challenging task of developing large export-oriented enterprises. Indeed, from the figures on the size structure of establishments given in Chapter 2, it is clear that Irish manufacturing possesses few large units even by reference to countries of comparable size. While establishments and enterprises are by no means synonymous, the Telesis Report (1982) took the view that few Irish indigenous manufacturing enterprises were large enough to have the internal capability of developing new products and penetrating new markets abroad. This, however, invites the question as to why, if there are no entrepreneurial deficiencies, the larger enterprises have not combined in one way or another — through mergers, group arrangements or whatever — to take advantage of the potential economies of scale in marketing and technological innovation?

A potentially relevant piece of information bearing on that question is the rate of return derived from different forms of investment in Ireland. Ruane and Neary (1982) calculated the following rates of return in 1980 to five different assets, taking account of inflation and taxation: 6.8 per cent for an owner-occupied house, -3.6 per cent on a long-term government security, -8.5 per cent on a building society deposit, -11.9 per cent on a bank deposit, and -14.4 per cent on equities. Such findings might be adduced to support the view that the problem is not one of lack of productive enterprise but rather of lack of incentives to evoke productive enterprise: in other words, the problem lies in an economically unrewarding environment. That conclusion, however, is open to the objection that the rewards to enterprise are not solely determined by the environment, but also by the ability of enterprise to exploit any given environment. The fact, already cited, that US companies operating in the same environment could earn massive profits, assuming that these profits do not mainly arise from transfer pricing, would be sufficient to demonstrate this. Moreover, the state has gone to considerable lengths in Ireland to improve the environment for manufacturing enterprise, especially when directed to export markets, through grants, zero or low taxation and other incentives — without attracting a powerful response from indigenous industry. Thus figures on low yields to manufacturing investment by Irish companies could at least equally plausibly be interpreted as demonstrating the poor quality of indigenous manufacturing enterprise.

The foregoing discussion is necessarily inconclusive and will remain so until much further research has been carried out. What is clear is that the record of indigenous manufacturing enterprise has been a poor one. What remains in doubt is the reason for this. There are three broad categories of explanation.[26] The first emphasises imperfections in the price structure (e.g., excessive wage rates, taxation etc.) which it is argued reduce the incentive to manufacturing enterprise. The second stresses more fundamental imperfections in the economic environment in the form of barriers precluding access to markets or to key factors of production (e.g., technology, finance) and which cannot be overcome by enterprises themselves without outside assistance even if the price structure were changed. The third explanation lays most weight on social-psychological influences, affecting either the responsiveness of entrepreneurs themselves or the environment in which they operate. These social/psychological factors need not be due to innate personality characteristics, but could arise from the country's traditions and experience. Tradition and experience, however, can influence not only the social/psychological attitudes but also the more "objective" entry barriers referred to in the second explanation. The competence and skills of entrepreneurs are often specific to particular activities, and may not easily be developed where such activities do not already exist.

None of the three possible explanations can be rejected on the evidence at present available. Accordingly it would seem prudent in the present state of knowledge to consider any feasible policy measures — whether they be in the economic, social or psychological domains — that have a reasonable prospect of improving the situation. Nevertheless, the evidence presented by O'Malley (1983) and O'Brien (1985) on the pervasive nature in small, peripheral, newly-industrialising countries of the second class of deficiencies mentioned in the previous paragraph — barriers to entry and expansion — suggests that policy measures to overcome these deficiencies offer the greatest prospect of progress in developing the indigenous industrial base.

3 *The Role of Small Industry in Entrepreneurial Development*

Does the development of small industry have a useful role to play in evoking or creating greater entrepreneurial activity? In fact much of the larger indigenous manufacturing enterprise in Ireland began as small firms in the protected era of the 1930s–1950s. Indeed, historically most countries have used tariff and quota protection in the early phases of industrialisation. The fact that they have done so, even though standard economic theory holds that in competitive conditions it is nearly always possible to devise a superior alternative (e.g., production sub-

[26]This is a slightly different classification of explanations from that given in O'Connor and Lyons (1982), who combine the first two above into one (economic) explanation, and divide the third as between the social environment and the individual psychological impulses.

sidies), suggests that something more than merely changing the price structure is involved. Protection can play a role not performed by more "efficient" methods, such as production subsidies, in assuring access to a market and in reducing uncertainty. Given the small size of the home market in Ireland, however, protectionism could only be of limited effectiveness, and is now almost entirely ruled out by our EEC obligations. But free trade, while on balance more favourable to the economy, may also make it more difficult for new enterprises to emerge, if — as experience generally in industrial development suggests — most new enterprises tend to begin by supplying the home market where familiarity, transport costs etc. facilitate the establishment of a competitive edge. This would create a general presumption in favour of special measures to encourage the emergence of new indigenous manufacturing industry, much of which initially will be small.

It is true, of course, that the vast majority of small enterprises in any country do not develop into large enterprises. Nevertheless, the existence of a large pool of small enterprises can help in entrepreneurial development generally. It disseminates knowledge about the requirements of enterprise in terms of profitability etc. It encourages initiative in those who might otherwise think only in terms of secure employment. If the only models of enterprise were branches of large multinationals, or long established domestic firms, the psychological barriers to be overcome by new entrants would be greater. Employees in small enterprises have been a more significant source of new businesses in Japan than those engaged in larger, more bureaucratic, firms (Watanabe, 1970). And, as mentioned in Chapter 3, small industry provides a relatively economic way of sifting the good from the bad, and provides guidance, rooted in actual performance, as to which enterprises might best be selected for larger scale development.

In making these points, however, we would not claim that the encouragement of new small enterprises in the private sector is the only, or perhaps even the main, avenue to the development of indigenous manufacturing entrepreneurship. The development of established private enterprise, the scope for new forms of enterprise, such as workers' or producers' co-operatives, and even the possibility of direct state manufacturing enterprise in selected areas, are also major avenues that should be considered. Moreover, in seeking to develop entrepreneurship, measures other than direct instruments can be of great importance: for instance, Fogarty (1973) and O'Connor and Lyons (1982) stress the major role that can be played in the long-run by the educational system, while Sweeney (1981) argues for a wide range of measures aimed at creating a "technical culture". Finally, even in relation to small industry itself, the focus should not be solely on the emergence of new enterprises but also on the crucial features of the transition from small to large. We will take up these issues again in discussing the policy implications of our study in Chapter 10.

Chapter 6

SMALL INDUSTRY POLICIES IN OTHER COUNTRIES

Before going on to consider small scale industrial development policy in Ireland in Part II, we examine in this chapter the kinds of policies which other OECD countries have adopted in relation to small industry. It has to be acknowledged at the outset that while the details of these policies are relatively easy to collect, it is often not possible to determine their scale, usage and effectiveness.

1 The General Nature of Policy Interventions

Countries differ considerably in their overall industrial policies, and within these in the extent to which they pay special attention to the problems of small industry. While we are concerned here primarily with the policies affecting small industry, it should be pointed out that some of the differences among countries in approach to small industries reflect more general differences in approach to overall industrial policy. Advocates of the efficacy of the market system tend to question any state industrial policy intervention, holding either that the market works well or that, even if it does not, state intervention will only make matters worse rather than better. Such a *laissez-faire* approach to small industry, however, does not characterise the present-day policy of any of the countries we examined.

Two contrasting patterns of intervention in favour of small industry can be identified — active neutrality and positive discrimination. Active neutrality consists in interventions designed to remove disadvantages facing small firms that are not justified by the operation of free market competitive forces. These may arise from legal, institutional or administrative factors, or from imperfections in the market place itself. In practice, of course, it is often difficult to determine whether disadvantages faced by small firms arise from genuine diseconomies of scale or from artificial discrimination. Thus, for example small industry may be faced with high bank interest rates partly due to the higher costs and risks associated with small loans and partly due to unduly conservative evaluation of risk by bankers. In principle, a policy of active neutrality would seek to redress the latter, but not the former, disadvantage.

Positive discrimination, on the other hand, would involve the provision of facilities and incentives to small firms which are not offered to large firms. The basis of such an approach may be that the market is sufficiently imperfect in relation to new small entrants that only by positive discrimination can such firms be put on an equal footing with large. Alternatively, it may be based on broader economic or social benefits that are thought to accrue from having a vibrant small industry sector.

The political philosophy as well as the economic environment helps to determine the balance between these two approaches adopted in different countries. The US and Japan are both characterised by positive discrimination in favour of small firms — but for rather different reasons. In the United States small businesses constitute a powerful political lobby, and the general ethos lays great stress on the preservation of a healthy small business sector on both economic and social grounds. This has resulted in the provision of a wide range of financial and advisory services for small firms that are not made available to large firms.

In Japan, attention was devoted to the problems of small firms as long ago as the 1920s, when emphasis was placed by policy-makers on defending small-medium entreprises from the pressures of large enterprises as well as from the effects of general depression. A prime emphasis in Japanese policy is the encouragement of small firms to secure the advantages of large ones through rationalisation, mergers or co-operative arrangements. The aim of policy has not been to aid small firms indiscriminately but to encourage selected projects to grow on the basis of their own strength and efficiency — though attention generally is directed to removing disadvantages facing small enterprises and to improving the socio-economic standing of small business employees.

In both the US and Japan, however, it would be wrong to conclude that such positive discrimination involves cosseting small firms or sheilding them from market forces. On the contrary, as we shall see, both the US and Japan are quite demanding in their expectations of small firms, and policy measures are geared to realising these expectations.

As the other end of the policy spectrum, the UK and West Germany have traditionally tended to avoid according special treatment to small firms. There are a large number of countries which, although they do not practise positive discrimination, offer extensive technical and financial aid to small firms to enable them to surmount barriers in specific problem areas. The need for discriminatory measures is less apparent in countries where small firms are well integrated with the rest of the economy. This would appear to be the case in some of the smaller European countries where small firms are not treated as a separate sector for which special policies are needed, but are directly concerned in general industrial policy measures.

While historically the majority of Western countries have followed a policy of active nuetrality, in more recent years, there has been a considerable broadening in the scope of policies and institutions aimed at helping small industry. In some countries, which traditionally operated a policy of either active neutrality or non-intervention by the State in relation to small enterprises, there has been a shift in the direction of discrimination in favour of small firms. Many new policy instruments are aimed at tackling specific barriers to the start-up and growth of small firms, such as measures to encourage technological innovation in new small firms through technical advice and new financial sources. The pressing need for more jobs, and the fact that many countries are experiencing a net decline in manufacturing employment in the large firm sector, has focused more attention on the employment-creating potentialities of small firms. The growth of regional policies has also accentuated the tendency towards positive discrimination. Small-scale industry can be used as an effective instrument of regional policy, especially in locations that could not sustain large-scale industry. Even in larger locations, dependence on a single large-scale plant can have traumatic social consequences in the event of failure. Moreover, the local availability of efficient subsuppliers is helpful in attracting larger industry to a particular location.

2 *Institutional Framework*

The institutional framework of public policy aimed at small industry differs from country to country,[27] but is strongly influenced by the balance struck between the two broad approaches outlined in the preceding section. Thus the two major countries that have favoured the approach of positive discrimination, the US and Japan, both have a body with comprehensive responsibility for attending to small firms. In the US, the Small Business Administration (SBA), established in 1953, acts not only as the executive agency of government to administer various programmes aimed at small industry, but also engages in mobilising political support for such programmes. In Japan, the Small and Medium Enterprise Agency, established in 1948 within the framework of the Ministry of International Trade and Industry, plays a pivotal role in the planning and implementation of policy for small firms. It should also be emphasised, however, that in both countries the upper size limit of firms catered for by these agencies is very much higher than in Ireland. Moreover, the very special relationships that exist in Japan between small and large firms, as discussed in Chapter 3, must be borne in mind in any comparisons with that country.

Many countries have a host of public agencies, which developed at different times, catering either for small firms specially, or dealing with them alongside large firms. Technical and advisory services have long been provided to firms of

[27]See, for example, OECD (1971a, 1971b and 1978).

all sizes by technological institutes in a number of continental European countries. In 1906, the Copenhagen Technological Institute was established to provide technical assistance to Danish industry. Similar institutions were set up in the Netherlands in 1907 and in Germany in 1921. Initially these institutions concentrated on the modernisation and rationalisation of industry through technical advice. Since the 1950s, however, existing institutions were expanded and new institutions set up to assist industry in general and small industry in particular to adapt to modern techniques.

The proliferation of agencies led to widespread concern about lack of co-ordination.[28] In the 1970s industrial policy has been reviewed in many countries to secure greater co-ordination of measures. In Canada, an Enterprise Development Programme was launched in 1977 to replace a number of individual assistance programmes. Under this programme, comprehensive assistance is offered to small and medium businesses, taking account of the various financial, technical and managerial requirements of projects. In France, in 1976, an action programme was launched by the government to step up and co-ordinate services and programmes fostering the development of small and medium enterprises, giving local chambers of commerce a greater role in providing technical and general advisory services to small businessmen. A feature of public policy in many countries has been a greater co-ordination of industry research and development with the work of universities and technical institutes, as well as greater liaison between management institutes, universities and businessmen themselves.

In addition to greater co-ordination of public aid, the administrative machinery of public bodies has been decentralised in many countries, so that the local and regional fieldworkers in these bodies have greater responsibility and more contact with local businessmen. The need for regionalisation of services is all the greater in countries where there is a large spread of small firms over a big national territory. In Sweden, for example, aid to small and medium-sized enterprises is mainly handled through Business Development Associations based on counties, which administer most government programmes and combine a wide range of services relating to finance, information and advice. In Canada, the Enterprise Development Programme is managed by regional boards in each of the provinces under the control of a central board. In the US the SBA, while retaining centralised decision-making on major issues in Washington, delegates a certain degree of authority to state offices of the SBA which are free to implement and modify programmes of aid to suit local needs.

Added impetus to regionalisation of programmes for small industry has been given by the fact that many governments have adopted policies to develop

[28]See, for example, Beesley and Wilson (1982) in regard to the situation in the UK.

industry in backward regions. Special measures to develop small industry are often a significant component of such policies. In the UK, for example, the regional development agencies in Scotland, Wales and Northern Ireland provide special financial and advisory services to small firms. Equity and loan finance are provided as well as guarantees on loans from other sources. Locally based advisers are employed by these agencies at a relatively low cost. Loans are provided on favourable terms for small enterprises to cover up to 80 per cent of the project cost, with repayment spread up to 20 years. The coverage of these schemes, however, is thin.

Since 1979 special measures have been implemented in Japan to assist small businesses clustered in regions particularly affected by recession. Small local businesses which are seriously affected by adverse business conditions may qualify for special industry status and thereby be entitled to preferential tax treatment, and special loans and grants. The schemes are administered by the Small and Medium Enterprise Agency and involve the local business associations in the formulation of plans for rationalising and strengthening firms.

3. *Finance*

Most countries have introduced special measures to improve the access of small firms to investment funds. In general, governments have sought to improve the flow of loan capital, and to a lesser extent equity and venture capital, to small firms through private financial institutions. Many governments have also established public financial institutions to supplement the facilities offered by private institutions. In some countries, governments have gone further and introduced schemes of grants for developing small firms. In general, the more generous the assistance provided from public funds, the more specific tends to be the purpose for which it is given (e.g., finance of innovations). Some use has also been made of taxation policy to encourage the flow of funds to small industry.

Loans and Loan Guarantees

The financing requirements of small firms differ according to the traditions of banking and types of company ownership in different countries. For example, in West Germany, many small firms are family owned and are, according to the OECD (1982) study, well catered for by local savings banks and community banks, and the venture capital market has remained underdeveloped. By contrast, in the US where there is a highly developed venture capital market as well as a tradition of a high company formation rate in small innovative firms, entrepreneurs in small firms are less reticent about diluting the firm's ownership by attracting external equity funds. The SBA can provide direct loans to small businesses (up to 1,500 employees) but, more important, it was authorised under legislation in 1958 to license Small Business Investment Companies (SBICs). These are private companies designed to take equity in small firms and to make long-term loans to them. The SBA may itself invest public funds in the SBICs, which are also accorded certain tax advantages.

In Canada, the Federal Development Bank, established in 1975, specialises in financing small businesses through loans, equity and leasing. The bank has also developed a broad range of services covering finance, management training and counselling. In Japan, small businesses seeking finance are catered for by the Small Business Finance Corporation, which specialises in lending to firms with less than 300 employees, and the People's Finance Corporation, which specialises in funding firms employing less than 50. Both institutions are government owned. Government policy towards small enterprise in Japan is focused on encouraging small firms to help themselves through technical advance and combination with other firms. Special low interest long-term loan facilities are offered by the public lending agencies to encourage small firms to modernise or strengthen their structure, often through inter-firm co-operation in a wide area of business activities. In certain industries, firms which employ 20 persons or less are eligible for 50 per cent financing of capital equipment through interest-free loans granted by the People's Finance Corporation. They are required to raise the remainder of the funds themselves, and to meet this need there is a publicly funded Equipment Leasing Agency which purchases machinery and equipment for leasing to small and medium firms.

Non-repayable cash grants towards fixed capital are not as widely used as a policy instrument in other countries compared to Ireland. Where grants are provided, they are normally either confined to depressed regions or are directed towards overcoming a specific cost barrier in an area such as R & D, innovation, or marketing. On the other hand, loan guarantee schemes are much more widely used abroad. Typically, loan guarantees are part of a package of assistance offered to small firms. Usually the public authorities themselves, or through their agencies, agree to act as guarantor of 80–90 per cent of the amounts borrowed through commercial banks or other lending agencies. In Japan, mutual guarantee associations of private firms have been established on a regional basis under the Credit Guarantee Association Law, and the associations in turn are guaranteed by the Small Business Credit Insurance Corporation, a wholly-owned government corporation.

In the UK, a Loan Guarantee Scheme administered by the Department of Trade and Industry was introduced in June 1981 on an experimental basis for a three-year period. The government underwrote 80 per cent of approved loans not exceeding Stg£75,000 made by commercial lending agencies. Only firms which failed to obtain finance from a financial institution due to insufficient security were eligible for a guarantee. The repayment period of the loan was five to seven years, and borrowers availing of the scheme were required to pay a premium of two and a half per cent on the government-guaranteed portion of the loan. Total government guarantees were initially limited to £50 million, but were eventually raised substantially as the demand exceeded expectations. In the period June 1981 to November 1983, there were over 12,000 loans amounting to over Stg£400 million or an average loan of Stg£35,000 (Stewart 1984). Defaults, however, amounted to approximately £40 million, and the scheme was altered to reduce the government guarantee to 70 per cent and increase the surcharge to three per cent. Loan approvals dropped considerably after these changes. The rationale of the surcharge has been questioned by Binks and Coyne (1983) on the grounds that it discourages demand in an undesirable way. The government's intention was that the scheme be self-financing, but this has the effect of making the successes under the scheme pay for the failures.

Member states in the EEC are eligible for assistance from the European Communities towards the financing of small and medium firms. The most important channel of assistance has been the European Investment Bank (EIB) established in 1958. The EIB extends loans to undertakings, public authorities or financial institutions for investment projects in all sectors of the economy in certain disadvantaged regions which meet the Community's priority aims. As a general rule, the EIB can only grant direct loans for sizeable projects, but small and medium-sized enterprises are specially catered for by a global loans system operated in conjunction with national agencies in member states. Under the system, loans may be granted for up to 50 per cent of the fixed investment costs: in terms of Irish pounds, the range is very broadly from a minimum of IR£15,000 to a maximum of IR£5 million. The criteria used to determine size eligibility of projects are net fixed assets, which must be no greater than about IR£20 million before the project is carried out. In some countries the national governments undertake to bear the rate of exchange risk on loans. Where funds are borrowed in strong currencies and lent to projects in countries with high inflation and a weak currency, the rate of interest on loans is somewhat below the average charged on domestic loans by financial institutions responsible for handling EIB loans. Funds for the financing of small and medium firms are also available under the New Community Investment (NCI) or Ortoli facility and are channelled through the European Investment Bank to "intermediate" financing bodies for on-lending to small and medium firms. NCI loans are available for financing projects in all areas of the EEC, and not just certain disadvantaged regions in each country as is the case with EIB global loans.[29]

[29]Further details of the EEC schemes are given in Commission of the European Communities (1984).

Interest rate subsidies, financed by the state, are used in several countries (e.g., Japan, Italy, Denmark and Australia) and may apply to loans offered by private as well as public agencies. In the US, the SBA provided direct loans at a statutory fixed rate of five and a half per cent. Since this later fell well below commercial rates and Congress limited the funds for the programme, the direct lending activities of the Agency were curtailed and it switched to guaranteeing loans made by other institutions at commercial rates. In Australia, there is a government imposed interest ceiling on trading bank loans of less than Aus$100,000. The Bureau of Economics (1981) has argued, however, that this has had a negative effect on small business development on the grounds that it led to rationing of funds, which bore relatively heavily on innovative firms.

Finance for Innovation

One of the main concerns of policy in relation to small industry in the last decade has been to ensure that access to funds for innovative firms is open on at least an equal footing to other firms. In West Germany, the Ministry of Economic Affairs offers grants of up to 50 per cent of the eligible cost of R & D pre-production projects, repayable within ten years if the project is commercially successful. This applies to all sizes of firms, however, and it has been estimated by Little (1977) that only two per cent of total R & D spending in 1973 in private industry was undertaken by firms employing less than 500. In the same year only 0.9 per cent of public funds towards R & D was allocated to firms employing less than 500. The proportion of public funds allocated to small and medium firms has been rising over time, however. Nevertheless, the proportion of all kinds of grant allocated to small and medium enterprises (employing less than 500) by the Ministry of Economic affairs was only nine per cent in the early 1980s (OECD, 1982).

The French government has been giving increasing attention since the mid-seventies to small and medium firms, especially new innovative firms. In the period 1978–79, 14 per cent of total funds allocated on public programmes of assistance in aid of R & D and innovation was allocated to small and medium firms (i.e., those employing less than 500). For both large and small projects, grants are available from public agencies to cover 50–75 per cent of research costs of projects and 50 per cent of the costs of development. Funds granted for the latter purpose are repayable at a fixed percentage rate of gross product sales, ranging from two to 20 per cent, if the project is commercially successful. In addition, the proceeds from the sale of licences or patents resulting from such successful innovation is shared with the government at a rate of 25–50 per cent (Rothwell and Zegveld, 1978).

In Belgium, grants are available from the Institute for the Promotion of Scientific Research in Industry and Agriculture (IRSIA) towards the cost of research. The proportion of total IRSIA grants which was allocated to small and medium firms was six per cent in 1975, rising to 15 per cent in 1979. In Italy, the Istituto Mobiliare Italiano provides grants of 40 per cent of the costs of applied research, repayable if projects are a success. In the early 1980s, almost one-half of aided projects were firms employing less than 300, and these accounted for 28 per cent of the total funds amounting to IR£60m (per annum). Under the Enterprise Development Programme in Canada, non-repayable cash grants are provided towards productivity improvement studies as well as for innovative projects involving new technology. Up to 50 per cent of the costs of employing specialised labour and special purpose equipment and materials is paid under this programme. In Britain the National Research Development Corporation (NRDC) was established in 1948 and was specially charged with the financing of invention in British firms. Between 1948 and 1976, the NRDC engaged in 930 joint venture investments with enterprises, scarcely a large number in a country the size of Britain. Of these 366 were commercially successful and yielded a royalty income to the NRDC. About 50 per cent of NRDC assisted firms in the 1970s employed less than 200 persons (OECD, 1982). The NRDC established a Small Company Innovation Fund. The NRDC itself has now been incorporated into a new organisation — the British Technology Group (BTG). The BTG sponsors enterprises to engage in research and development either on their own or under licence from the BTG. In certain cases, the BTG extends non-repayable grants.

Finance for Marketing

Guarantees on loans for exporting firms are widespread. For example, in Italy the Instituto Centrale di Credito a Medio Termine provides insurance against the commercial and political risks attaching to short-term credit. The Canadian Export Development Corporation also provides insurance against non-repayment by foreign buyers. In the UK, an Export Market Entry Guarantee Scheme was launched in 1977, under which loans are available towards 50 per cent of the cost of new export market ventures. The loan is repaid at the commercial interest rate out of the receipts from new export sales. These schemes are available to all firms regardless of size. Since small firms export relatively less than large, the latter tend in practice to secure a relatively greater share of such loans and guarantees. For those small firms which do benefit, however, the gain is probably greater since small firms generally are more vulnerable to market failures and bad debts.

Taxation

Taxation provisions provide a policy instrument for influencing the flow of funds to enterprises of different sizes as well as encouraging certain types of activities, e.g., R & D within firms. The impact of taxation measures is particularly difficult to quantify. Tax incentives generally only affect profitable enterprises directly and may, therefore, exclude many new enterprises which experience difficulties in earning profits in the early stages of an enterprise.

In some countries the rate of company profits tax is graduated according to the size of taxable profits. In the UK the corporate tax rate on small companies (those with profits up to £100,000 per annum) was reduced in 1984 from 38 per cent to 30 per cent. The 38 per cent rate compared with a maximum of 52 per cent, but the maximum rate is to be cut progressively to 35 per cent from April 1986 onwards. In Japan, in 1981, for companies with less than 100 million yen in capital, the tax rate was 28 per cent on profits up to seven million yen and 40 per cent on larger amounts. For companies exceeding 100 million yen in capital, a rate of 40 per cent applied on all taxable profits. In the US the rate of corporate tax rises from 17 per cent on the first $25,000 to 46 per cent on $100,000 or more.

Virtually all countries provide capital allowances in respect of new investment or depreciation, but these allowances are usually available to all firms regardless of size. Likewise, R & D expenditure is usually allowable against profits for the purpose of calculating liability to tax for all firms regardless of size.

Tax measures have been introduced in some countries to encourage individuals setting up new enterprises. In the UK, the 1981 Budget provided for tax relief of up to £10,000 for investment in *new* business, a scheme referred to as the Business Start-up Scheme. The Budgets of 1982 and 1983 raised the amount of tax relief, bringing it up to £40,000 in 1983. The Business Start-up Scheme was not successful, containing many restrictions to prevent abuse. The 1983 Budget extended the relief to investment in established unquoted companies, and is now known as the Business Expansion Scheme. The take-up on the Business Expansion Scheme is greater, but so also is the amount of abuse.

In France, there is an allowance of 33⅓ per cent on taxable profits in the first year of a company formation and in the three subsequent years. In Denmark, expenditure on innovation in the pre-commercial period of a project is depreciable in the first commercial year. The acquisition of business premises may be a significant cost penalty for small firms in most countries, especially where a firm is located in an urban area.

Some countries have used tax as well as other financial incentives to encourage small enterprises to co-operate more closely. In Japan, which has gone furthest in this direction, businesses belonging to a trade association are entitled to additional tax relief on depreciation over and above their ordinary depreciation allowance on fixed assets designed to modernise the firm. Similarly, businesses are relieved of tax on reserves which are transferred to recognised associations for joint use. Payments made to associations are deductible for tax by both the associations and the constituent businesses.

4 Technological, Marketing and Advisory Services

Countries differ considerably in the extent of government involvement in fostering technological change in small industry. Japan is a good example of a country with a well-developed public technological infrastructure benefiting all firms including small. There is a large spread of public laboratories throughout the country which provide assistance through counselling, testing of products and processes, and dissemination of technical know-how. Most of these laboratories have been in existence for 30 years or more. Since the 1960s, facilities have been extended to businessmen to test products and materials. The laboratories are co-ordinated by the Small and Medium Enterprise Agency. The Small Business Promotion Corporation (SBPC) — a financing agency — has responsibility for the training of technical personnel employed in public laboratories as well as personnel in small firms. If small firms are in receipt of government subsidies towards research, and if they avail of technical courses provided by the SBPC, they are required to disclose their research findings at these courses. The SBPC also facilitates technology transfer from large to small enterprises.

In the USA, government support for scientific and technical development is mainly chanelled through the National Science Foundation (NSF). In 1980, small and medium enterprises in the US (i.e., those employing less than 500 persons) accounted for less than 10 per cent of the total value of federal R & D contracts going to industry. Moreover, in spite of the high innovativeness of US small companies (National Science Foundation, 1977), R & D spending is heavily concentrated in large companies. The NSF has sponsored the setting up of Innovation Centres on university campuses, to provide formal education and clinical experience in invention, application of new ideas to productive processes, and entrepreneurship. Independent inventors as well as university students may avail of these services. Feasibility study grants are available from the NSF for small businesses. The grants are staged to enable the project to be monitored effectively throughout its development.

In Denmark, the National Agency for Technology (NAT) provides technical assistance for all sizes of firms along with a number of technical institutes and universities. According to the OECD (1982) study, inter-firm co-operation in R & D is comparatively limited in Denmark due to the proximity of markets and customers to the source of production and the consequent higher level of competition. A system of "service centres" supported by NAT provides technical information in various parts of the country. A Danish Invention Centre was established in 1972. By 1978, the centre was dealing with 900 inventions per year. The activities of the centre encompass advice, evaluation, workshop facilities for testing and development, and facilities to promote licence contracts. Only about two per cent of inventions processed in the mid-1970s, however, actually reached product state (Rothwell and Zegveld, 1978).

In France, specialised regional technology centres exist to assist firms in R & D and innovation. These centres are located in a particular region and normally specialise in a certain product range typical of the region. Regional Agencies for Scientific and Technical Information (ARIST) are managed by chambers of commerce and act as information centres. These agencies have been extended to all areas of France in recent years. Industrial Creation Centres also exist in the regions to provide facilities for entrepreneurs to test and develop new products or processes. France is no exception in providing more regionalised technology services. Most countries have recognised the importance of an adequate technology infrastructure sufficiently accessible to the producer/innovator.

Marketing

In regard to the home market, government action in many countries is limited to giving information about purchasers and advice on quality and design. In some countries, however, care is taken that small firms are put, as far as possible, on an equal footing with large firms in bidding for government purchases. This is particularly so in the United States where the SBA has power to enforce procurement arrangements designed to ensure that small firms secure a "fair proportion" of government orders, contracts and subcontracts. Departments can be required to confine bidding for specific contracts to small firms, and more generally to divide contracts into the smallest economic lot sizes so that small firms can bid for them. Where a small firm's bid is rejected because of doubts about its ability to complete the contract, the SBA may, after investigation, issue a "certificate of competency" which requires the department to accept the bid, with the SBA assuming responsibility for non-performance. In Germany also, tenders for public authority purchases are distributed in lots sufficiently small to facilitate participation by small and medium enterprises, while group bids are encouraged to enable small firms to co-operate together.

State export boards exist in one form or another in most countries, providing a varying range of services to firms of all sizes — credit guarantees, market research, organisation of trade fairs etc. As pointed out in Chapter 3, however, Japan is one of the few countries where small firms make a substantial contribution to exports. A large proportion of external trade for all sizes of firms is handled by trading companies. Although these companies cater for all sizes of firms and are not state bodies, it may, nevertheless, be of interest to describe their activities briefly since they are, *inter alia*, important to the viability of the very extensive small industry sector in Japan.

The trading companies are made up of affiliated companies and act as trade intermediaries: preparing documentation for exporting and importing, obtaining trade finance, arranging insurance and transport and providing for storage. They also provide information on markets to both overseas and Japanese companies. In 1980, there were over 8,500 trading companies in Japan handling 62.8 per cent of all export trade and 64.6 per cent of all import trade. The nine largest trading companies accounted for 40.1 per cent and 47.4 per cent, respectively, of export and import trade. The problems of language, differences in customs and complexities of distribution in most foreign markets require a minimum level of commitment and investment not normally within the capacity of small enterprises. Small enterprises, in particular, stand to benefit from the services of trading companies by sharing in the advantages of economies of scale in buying and selling, as well as in the spreading of risks. Many trading companies also provide trade credit as well as other forms of financial aid to affiliated companies, including equity investment, direct loans and guarantees for loans. Because of their closeness to suppliers and customers, trading companies are often in a better position to judge the credit-worthiness of their client companies than banks. Since the 1960s, many trading companies have extended their role as trade intermediaries to include direct investment in other companies as well as direct involvement in manufacturing.

The general trading companies in Japan tend to be large in size and handle a wide variety of products. These companies may, therefore, finance losses in one activity with the profits in other product areas. Similarly, if exchange rates are altered, trading companies are in a position to absorb some or all of the losses incurred, either on exports or on imports. This arises due to the fact that they are trading both imports and exports through contracts denominated in a major international currency. In addition to the large general trading companies, there is a great number of small trading companies, many of which specialise in areas where major trading companies experience difficulty in competing. Many of the smaller, more specialised trading companies have become active in handling goods which are transported in smaller volume or which require extensive marketing and associated servicing after delivery.

A growing area of activity in recent years has been the involvement of trading companies in trade among countries not involving Japan. Through their network of distribution centres, trading companies have been utilised by overseas manufacturers for the purpose of finding markets not only in Japan but in other countries as well. Trading companies and other manufacturers are aided by the official Japan External Trade Organisation (JETRO). JETRO provides marketing aids and services which include market information and research, as well as the organisation of trade exhibitions and fairs. Over the last two decades, JETRO has evolved from being primarily an export board to a situation now where its handling of import trade is as important as export trade. It advises overseas manufacturers on market opportunities in Japan. The most efficient sources of overseas supplies of raw materials, energy and components for processing is an important consideration for many Japanese companies.

General Advisory and Management Services

In many countries, public agencies provide a variety of services including advice, training of personnel and management consultancy in all areas of running a small business. In the UK, as already mentioned, the Department of Trade and Industry operates a Small Firms Information Service which consists of a network of information centres located throughout the UK. These centres provide information free of charge from trained employees by telephone, face-to-face, or by pamphlet. A service introduced in the UK in the last few years gives access to counselling by retired executives and others. The first hour's counselling is free, but thereafter a fee is charged. The counselling service is oversubscribed, probably because it provides direct "on site" advice which small firms often need most. A similar service, known as the Counselling Assistance for Small Enterprise (CASE) scheme, is run by the Federal Business Development Bank in Canada. Under the scheme, management consultancy specifically geared to firms employing less than 75 persons is provided by retired businessmen at rates subsidised by the government. In 1981, about 14,000 clients availed of CASE compared to 2,800 in 1976.

The idea of using retired businessmen and executives as counsellors to small industry was first developed in the US in the 1960s. The SBA launched the SCORE programme, an acronym for Service Corps of Retired Executives on a nationwide basis in 1964. SCORE counsellors include businessmen, engineers, bankers and lawyers. There were 6,000 SCORE volunteers in 1976. No fee is charged if the counsellor travels less than 25 miles to visit a small business. The consultancy service of SCORE is augmented by the Active Corps of Executives (ACE), whose members are still active in business, industry and in educational institutions and numbered 2,500 in 1976. Counselling services cover all aspects of managing a business from raising capital, filling in application forms, marketing or technical change.

In Israel, the Productivity Institute provides an extension service for small firms akin to that provided for Irish farmers by the agricultural advisers. Rather than waiting for firms to seek information and advice, the adviser calls on the firm, discusses its problems, and even if unable to help directly is in a position to pin-point for the firm the source of help. For this purpose the advisers are trained in a wide range of business knowledge — technical, marketing, financial, managerial and personnel — relevant to the position of small firms.

Greater university-industry liaison has been promoted by policy-makers in most countries. In the US, the Small Business Institutes provide a counselling service to SBA clients by means of senior and graduate business students. No fee is charged for this service. The SBA also runs a Small Business Management Development Programme in co-sponsorship with universities, colleges and local business organisations. Problem clinics and pre-business workshops are organised. In France, Small and Medium-sized Enterprise Institutes were set up in 1979. These institutes provide free consultancy services through selected business students. Small Business Centres exist in a number of universities and polytechnics in the UK. In addition to providing consultancy services and courses, these centres carry out research related to small industry.

The use of local chambers of commerce in counselling and aiding firms is widespread in the continental European countries, where chambers of commerce generally have statutory status. In Germany and Austria, all firms are required by law to belong to chambers of commerce. This arrangement may be advantageous for small firms since in the absence of such involvement, small firms may remain isolated and unrepresented. In France, local chambers of commerce have fulfilled an extended advisory role since 1976. In Greece, chambers of commerce train managers, provide information and research findings, advise the state on business problems and promote exports abroad.

5 Conclusions

From the foregoing review of policy in regard to small industry, a number of general points emerge. First, it is clear that none of the countries considered adopts a completely "hands off" approach in relation to industrial development, even though all the countries are more highly industrialised than Ireland.

Second, all countries take at least some measures aimed specifically to help small industry, even if these measures are professed to represent no more than the removal of artificial disincentives facing small firms. Indeed it is noteworthy that the two major countries whose industrial capability is now most admired, the US and Japan, explicitly adopt an approach of positive discrimination in favour of small firms. As mentioned already, however, such positive discrimination is emphatically not a lame duck policy designed to shelter small firms from market forces.

In the US, small firms are looked to for the creation of more competition in an already highly competitive environment. There is a premium on rapid growth, and proprietors are expected to dilute their equity in order to finance that growth. In Japan there is a strong emphasis on flexibility and adaptation, and small firms are expected to co-operate with each other and with larger firms in achieving these goals. Indeed there is a sense in which the small firms sector is "exploited" in that it is expected to bear the brunt of responding flexibly to recessions and other economic shocks — a factor that may explain why, as mentioned in Chapter 4, wages are much lower and conditions of employment much less secure in Japanese small firms relative to larger firms than in Western countries. It is impossible to say, however, how much of the acknowledged vibrancy of small industry in the US and Japan is due to policy measures, and how much is due to the special conditions and ethos of those societies, of which policy itself may be a reflection.

A third general conclusion is that in most countries there has been an increasing degree of policy intervention to help small firms. A number of factors have given rise to this. Increasing unemployment has induced policy-makers to look to small firms as sources of job creation. Concern for peripheral or depressed regions, which may not provide attractive locations for large industries, has led to greater encouragement for developing more small industries in these regions. The technological backwardness of the generality of small firms, and the innovative potential of a minority, have both contributed to increasing the technological services offered to small firms.

Finally, in many countries, programmes aimed specifically at small firms have emerged piecemeal in response to perceived needs, rather than on the basis of a well-thought-out plan. This has often resulted in a proliferation of agencies dealing with small firms without adequate consideration of their effects and without clearly-defined overall objectives.

Apart from these general conclusions, some of the specific measures adopted in other countries might be considered for use in Ireland. These will be taken up again in the final chapter dealing with the policy implications of this study for the development of small industry in Ireland.

PART II

THE SMALL INDUSTRY PROGRAMME (SIP)

Chapter 7

SMALL-SCALE INDUSTRIAL DEVELOPMENT POLICY IN IRELAND

Since the early 1930s Ireland has had an active policy for developing manufacturing industry generally. In the 'thirties and 'forties the thrust of this policy was protectionist, and the major policy instruments were tariffs and quotas. These were supplemented to some degree by a limited amount of state enterprise in manufacturing activities which private capital did not enter, and by efforts to improve access to capital and to raise efficiency. In the early 'fifties the first steps were taken towards evolving an export-oriented development strategy, which has in the interval been extended and elaborated to a very great degree, and which is now in the course of re-formulation.[30]

Although no special meausres of any consequence were introduced specially for the benefit of small industry until 1967, in practice the operation of protection tended to favour small-scale industry, given the rates and speed at which it was introduced and the severe limitations of the home market. The Small Industry Programme (SIP) launched in 1967 remains the chief instrument of policy directed to small-scale manufacturing, and Part II of this study is mainly concerned with examining the operation and effects of that programme. In the present chapter, we first describe the SIP and how it evolved. For completeness, however, we also give a brief account of the other aids available to small manufacturing firms — some of them primarily confined to small, while others apply to small as part of the general industrial strategy.

1 The Origin and Scope of the Small Industry Programme

In October 1966 the Minister for Industry and Commerce announced a programme to assist small manufacturing industries in selected pilot areas,

[30]For accounts of the protectionist phase, see Ryan (1949) and Kennedy (1971). For the impact of industrial development policy since the 'fifties, see Survey of Grant-Aided Industry (1967), McAleese (1977), Telesis Report (1982) and O'Malley (1983). The new direction of policy is set out in the Government White Paper (1984) on Industrial Policy, which will be considered further in Chapter 10.

Carlow-Kilkenny, Limerick-Clare and Leitrim-Roscommon-Sligo The programme was launched in April 1967 and a special Small Industries Division was established within the Industrial Development Authority (IDA), which was at that time an industrial promotional organisation. The programme was operated in co-operation with An Foras Tionscail, which was then the body responsible for giving grants. Following the Industrial Development Act 1969, these two bodies were amalgamated with effect from 1 April 1970 to form the new IDA, combining both the promotional and grant-giving functions for manufacturing generally, including the SIP.

The objective of the programme was to assist in the development of existing small industries and the establishment of new ones as part of the overall programme of industrial development. It was felt that the contribution of small industries to the expansion of manufacturing output and employment would not be adequately realised without concentrating specifically on their special problems and according them somewhat more generous incentives and more intensive care. The programme was also seen as important for regional policy in that small firms could operate more readily than large in smaller towns and villages. Furthermore, it was intended that it would help in developing an experienced industrial workforce and in creating a favourable environment for enterprise generally (O'Connor, 1969).

In September 1967 the programme was extended to incorporate all designated areas (mostly in the poorer, western half of the country). By January 1968 the programme covered 15 counties, including those in the designated areas. In April 1969, the programme was further extended to include all areas of the country except Dublin city. The exclusion of Dublin arose from the regional policy goals which at that time aimed at a more decentralised development of manufacturing. In October 1975, however, Dublin was included for certain products (mainly engineering, plastic and chemicals) and in 1977 for all products except beef, dairy and related products. The emphasis of regional policy had shifted somewhat, and indeed the decline in manufacturing employment in inner city areas of Dublin has now become a matter of concern to policymakers. In an attempt to revitalise these areas, the IDA has constructed small industry centres in a number of inner city areas where large industry is reluctant to locate. Small firms are encouraged to establish in Dublin under the ownership of people living in Dublin. It is intended that these firms would not undermine the potential of projects in the rest of the country. The IDA now also supports community sponsored projects in Dublin's inner city as well as in other areas.

Administration of the SIP

The IDA administers the SIP in all areas of the country except for indigenous companies in the mid-west region (for which the Shannon Free Airport

Development Company, SFADCo.,[31] has had responsibility since 1978), as well as the south and south-west areas of Co. Offaly (for which SFADCo. has been responsible since 1980), while various areas of the Gaeltacht where the Irish language is spoken are dealt with under separate programmes operated by Údarás na Gaeltachta (see Section 2 re Gaeltacht areas). The incentives offered by SFADCo are similar to those provided by the IDA to small firms, and the cost of grants and other financial incentives to small industry projects approved by SFADCo is defrayed from IDA funds. SFADCo was given a mandate to promote small industry in a more intensive manner than had previously been attempted, in order to explore fully the potential of small industry in regard to employment growth and export activity. It was also asked to pioneer more intensive support systems for small industry on an experimental basis which, if successful, could be applied nationally.

For the purposes of qualifying for assistance under the SIP, firms were originally required to be manufacturing establishments employing 30 persons or less (or 50 in designated areas) and to have a written down value of fixed assets of not more than £60,000 (£100,000 in designated areas). Subsequently, these limits have been adjusted upwards, and the employment level prevailing at the present time is 50 workers (in all areas). A small number of firms engaged in service activities have also been aided under the SIP. Where a proposal for a new small establishment comes from a larger firm, it will only be considered under the SIP in the case of a product that is new to the firm and that does not involve vertical integration.

The administration by the IDA of the SIP is distinguished from other IDA programmes in that the financial incentives are more favourable, the range of services are more extensive and they are operated more intensively. As in the case of the major IDA programme, the New Industries Programme (NIP), the chief financial incentive is in the form of non-repayable cash grants towards the cost of sites, new buildings, the extension or reconstruction of existing buildings and new machinery and equipment. These grants are negotiable up to a limit of 60 per cent of the value of fixed asset investment in designated areas, and 45 per cent in all other areas, except Dublin,[32] where the maximum grant on site and buildings is 45 per cent and 35 per cent for plant and equipment.

Capital grants under the NIP are also negotiable up to a limit of 60 per cent of fixed assets in designated areas and 45 per cent elsewhere. Since the early 1970s, however, the IDA has operated administrative limits within these overall limits. The administrative limit is 50 per cent in designated areas and 35 per cent in non-designated areas. In practice, as we shall see later, grants form a much

[31]See Callinan (1984) for an account of the evolution of the role of SFADCo.
[32]The inner city of Dublin is at present a designated area and therefore eligible for higher grants.

higher proportion of fixed assets in SIP projects than in NIP projects, though because of the much lower capital intensity of SIP projects, the grant cost per job is less.

Due to the capital costs involved, even when reduced by grants, some small industrialists prefer to rent premises — at least in the early stages of the business. In such cases the IDA provides rent subsidies, while leasing grants can be made for plant and machinery. In addition the IDA has operated an advance factory construction programme since 1971 to provide factories of various sizes, under which over 500 small industry units had been completed by end 1983. A programme of building clusters of small factory units was first initiated on a pilot basis in 1974 in response to an apparent shortage of suitable premises for small industry. The IDA Industrial Plan 1978–82 provided for an accelerated programme of development. By end 1982, 53 such centres (including craft centres) were completed and a further eight were under construction or planned. The centres are designed to enable projects as they grow to acquire more floor space. Advance factories are normally rented on a 25-year basis with an option to purchase. The majority of SIP firms, however, are located in premises built or renovated by the promoter with the aid of an IDA grant. For this purpose, serviced industrial sites are provided in most towns through the IDA or the local authority.

In regard to working capital, special arrangements have been made between the Associated Banks and the IDA under which loans for working capital beyond those permitted by normal banking facilities but subject to a maximum, may be made available for SIP firms on a term loan basis for periods of up to five years. Grants are available for the training of workers (in liaison with AnCO, the Industrial Training Authority) and managers, and for consultancy services.

Processing of Proposals

Applicants for SIP assistance often have considerable difficulty in formulating proposals. While these difficulties may sometimes arise from no more than inexperience in the necessary paper-work, they also often reflect more substantive problems that require to be overcome before the proposal can be considered viable. Consequently the administration of SIP project proposals and subsequent follow-up tends to be far more staff-intensive relative to the size of the projects than in the case of other IDA programmes.

Moreover, relatively more of the work involved has always taken place at local level. While up to recently the applications were ultimately evaluated and decided on centrally, the County Development Officers and County Development Teams were closely involved. County Development Officers were originally full-time officers of the Department of Finance appointed by the government in 1966 to promote economic development in western counties. They were subsequently appointed in all other counties except Dublin. County Development Teams were set up in 1965 and involved in the promotion of the SIP from its beginning. These teams are composed of the County Manager, the County Engineer, the Chief Agricultural Officer, the Chief Executive Officer of the Vocational Education Committee, the Chairman of the County Council, the County Development Officer and the IDA Regional Manager. (The IDA has offices in each of nine planning regions into which the country is divided.) All County Development Officers in counties covered by designated areas are officers of the Department of Finance. The remaining development officers are engaged as staff of their respective local authorities. In 1970, the government decided that all applications for SIP grants had to be initially channelled through the local County Development Officer in areas where such an officer had been appointed.

The County Development Officers therefore acted as liaison between the project promoter and the Small Industries Division (SID) of the IDA, their role being to help the promoter with advice and, indeed, to seek out and encourage prospective promoters. A completed project proposal would be examined by a project executive of the SID, who would have further discussions with the promoter on the basis of his own evaluation of the project and any technical, marketing or financial advice he thought necessary to seek. Further adjustments in the proposal could follow, and if the project executive then judged the proposal to be satisfactory he would recommend it for approval to the Small Industries Committee, chaired by an executive director of the IDA. From 1982, each of the regional offices of the IDA was assigned a small industry specialist with authority to process and recommend projects, involving grants of up to £35,000, to the Small Industries Committee. A further step in the direction of regionalisation was the decentralisation to the Cork regional office of all the functions related to the SIP for Cork city and east Cork. The Government White Paper (1984) announced that the SIP will be fully regionalised, with primary responsibility for administering the programme being devolved to the regional offices.

The evaluation of a project involves such matters as the market potential, skill content, potential linkages with other firms, use of domestic raw materials, job numbers expected, as well as the financial arrangements. In regard to markets, no problem arises where the output is to be exported or will clearly substitute for imports. Where, however, the home market is already catered for by domestic manufacturers, assistance is limited to existing firms, and regional factors may be decisive in the evaluation. Particular difficulties arise in cases where there is a combination of import growth and under-utilisation of productive capacity. Similar difficulties arise in the case of sectors depending on native raw materials that are already in scarce supply. The IDA maintains a list of "sensitive industrial sectors" in which there is over-capacity or shortage of raw materials and when grant assistance is either not available or limited. In general, however, the policy is to examine each case on its merits and, where there is any reasonable scope for increased national activity, to give the promoter the benefit of the doubt.

After-care services are regarded as important to the successful outcome of a project. Contact is maintained through the County Development Officers and the regional officers of the IDA. There are regular visits to the firm during the initial two years following the first grant, especially in the case of new firms. In addition to identifying problems, recommending solutions and providing feed-back to the IDA, these visits are used to discuss opportunities for further expansion. In many instances approvals made under the SIP are given in stages, so that a number of years may elapse between the initial approval and the final payment. Following first grant approval, a firm may seek further financial assistance. If so, a follow-up report is generally made on the outcome of the first grant approval made, and if the IDA are satisfied with this report, a further grant amount is approved, and is treated as the second stage of an approval. Grant assistance under the NIP tends to be less staged than under the SIP. The IDA prefer to observe the initial impact of a grant approval for small firms before going on to make further approvals. In the case of new first-time projects, the promoter may be encouraged to start on his own in anticipation that if satisfactory progress is being made, assistance will then be approved. This approach is open to the criticism that the assistance is not available when most needed. Against this, however, must be weighed the risks involved in allocating public funds to wholly untried ventures. Also it would be misleading to imply that in such cases the project could have begun anyway: the prospect that assistance will be quickly forthcoming in response to initial independent effort is often crucial to evoking that effort.

Apart from processing projects and providing after-care, the SID has increasingly become involved in generating project proposals. In recent years, nationwide promotional programmes have been organised to encourage new entrepreneurship. Seminars, workshops and promotional evenings are held at various locations around the country. From 1975 a Project Identification Unit in the IDA helped firms to identify opportunities for supplying new products or engaging in import substitution, particularly in meeting the requirements for components and sub-supplies by other Irish manufacturers. Up to the end of 1982, 174 projects had been approved as a result of these investigations. With decentralisation the Unit no longer operates separately, the service being provided by regional officers in their own areas. The development of craft industries is dealt with under the SIP, in consultation with the Crafts Council of Ireland, an independent body funded by the IDA and acting as the main national agency for disseminating information and co-ordinating the efforts of local craft associations. An Enterprise Centre was opened by the IDA in Dublin in 1981 and another is under construction in Cork. The new centre aims to provide support services relating to management, marketing and administration for firms located at the centre.

To foster linkages with large industry, SFADCo launched a Matchmaker Service in 1981. The service promotes the sale of produce from small firms in the region and tries to act as a broker between small firms in the mid-west and large firms throughout the country. The Irish Goods Council (IGC), the Institute for Industrial Research and Standards (IIRS) and the IDA have services with similar objectives. The National Linkage Programme outlined in the Government White Paper (1984) aims to integrate these services. SFADCo maintains a permanent exhibition of small industry products in Limerick city.

The Innovation Centre in Limerick, organised by SFADCo in conjunction with the IIRS, IDA, the local National Institute for Higher Education (NIHE) and the National Board for Science and Technology (NBST), provides a product development service for small industry and training in entrepreneurship. There are facilities for "incubator" factories where new products and ideas can be tested and developed. In 1981 SFADCo in conjunction with NIHE-Limerick, the IDA and the NBST established a Microelectronics Applications Centre to encourage the use of microelectronic technology in all sectors of Irish industry. The purpose of the Centre is to develop new products and processes using microelectronic technology: it is not engaged in the design of microelectronic components *per se*, a function which is carried out at the National Microelectronic Research Centre in University College Cork. Both in the Innovation Centre and the Microelectronics Applications Centre have a national mandate.

Related Programmes

SIP firms which have expanded beyond 50 persons engaged become eligible for NIP grants and assistance in respect of further major expansion. Generally the administration of such grants and assistance is transferred from the Small Industries Division to the New Industries Division of the IDA but inevitably there is some overlap at the margin. In tracing the progress of employment in SIP firms later, we include all firms which received a SIP grant even if they later qualified for a NIP grant.

The IDA operated a Re-equipment Programme from 1968 to 1982, aimed mainly at established indigenous firms. Under this scheme, grants were made for fixed asset investment related to the modernisation or re-equipment of firms. Grants were limited to 35 per cent of fixed assets in designated areas and 25 per cent in all other areas. Re-equipment grants do not necessarily envisage any additions to the firm's level of employment. By contrast, the vast majority of grants made under the SIP are for projects where there is an anticipated increase in employment, whether in the expansion of existing firms or in the creation of new ones. The re-equipment programme was suspended early in 1982, except for commitments already made, and the Government White Paper (1984) confirmed the termination of the programme.

Applications for re-equipment grants for firms with less than 50 persons engaged were handled by the SID, though the grants were not treated as part of the SIP. We follow that distinction in this study, though it is arguable that it is of more administrative than of substantive significance. Some small firms received a re-equipment grant without getting a SIP grant, while others received grants under both headings. The former include firms located in Dublin which were ineligible for a SIP grant up to 1978, but which otherwise might have been classified as SIP firms. While the administrative classifications are by no means ideal from an analytical point of view, the seriousness of the problem is mitigated by the fact that the total volume of re-equipment grants for small firms has been small relative to the SIP programme.

There may also be an arbitrary element in treating another programme, the Enterprise Development Programme (EDP), separately from the SIP, though there are significant differences between the two programmes. The EDP, inaugurated in 1978 and administered separately from the SIP, is confined to first-time entrepreneurs from a segment of the potential business community which might not otherwise have sought aid from the IDA to set up a new enterprise. Persons who have qualified include professional, managerial and technical staff previously working in semi-state or private enterprises, as well as Irish people who were working abroad in technological and business environments. The expected employment in the first stage of an EDP project is typically of the order of 40–50 persons as against 8–10 in the typical SIP project. About half of the EDP projects are in electronics or other high-skill engineering activities,[33] and only a few in "traditional" activities like clothing or textiles, and even then only where they have special fashion/design characteristics. More than half of the EDP projects are aimed at least partially at export markets from the start, whereas the vast bulk of SIP projects are aimed initially at the home market. Under the EDP programme the normal financial and advisory assistance provided under the SIP has been supplemented by guaranteed loans for working capital purposes, interest subsidies, and in some cases direct equity shareholding where this is deemed useful.

Small firms, like large ones, can qualify for grants from the R & D programme administered by the IDA. Under the programme, which was first introduced in

[33]Cogan and McGovern (1984), however, found that "the number of genuine technology-based companies assisted under the EDP has been small".

1969, grants can be made available towards the cost of salaries, materials, prototype manufacture, testing and consultancy fees incurred in developing a new product or process. These grants are presently available at up to 50 per cent of eligible expenditure, subject to a maximum per project. In addition, the capital cost of permanent R & D facilities can be grant-aided subject to an upper cost limit. The R & D survey of industry relating to 1982, conducted by the National Board for Science and Technology, indicates that about 14 per cent of all R & D expenditure in manufacturing was contributed by the state. Most of this came from the IDA's scheme but there are other state schemes, such as the NBST Higher Education/Industry Co-operation Grants scheme. Grant assistance is also offered by the IDA to help individuals, community groups or firms to undertake feasibility studies into new manufacturing projects or to examine the possibility of developing new products in existing facilities. The scheme, which was originally confined to companies, was broadened in 1981 to include individuals and community groups. Up to 50 per cent of eligible expenditure can be covered by grant assitance subject to a maximum of £15,000 per study.

An indication of the relative scale of the major programmes is given in Table 7.1, which shows total grant expenditure in 1972, 1977, 1982 and 1983. The SIP has been one of the most rapidly growing programmes in this period, and in 1982 accounted for 18 per cent of total IDA capital expenditure but fell back in 1983 to 14 per cent. Of the total expenditure in 1982 and 1983 on the SIP, 74 per cent was in respect of the purchase of fixed assets, which was considerably higher than the figure of 62 per cent for the NIP. The difference is chiefly accounted for by the much higher proportion of NIP expenditure on grants for (a) leasing

Table 7.1: *IDA capital expenditure on industrial programmes, various years, current prices*

Programme	*1972*[a]	*1977*	*1982*	*1983*
	£m	£m	£m	£m
New Industry	16.0	32.6	88.1	102.8
Small Industry	0.8	2.5	25.1	21.9
Re-equipment[b]	6.6	11.0	10.6	8.3
Enterprise Development	—	—	4.6	4.1
R & D	0.1	0.4	3.3	5.8
Other	3.2	1.7	9.6	13.0
Total	26.7	48.2	141.3	156.0

a. Relates to year ended 31 March 1973.
b. Includes adaptation grants.
Source: IDA Annual Reports

machinery (16 per cent for the NIP as against four per cent for the SIP), and (b) interest subsidies (five per cent of the NIP total and negligible in the case of the SIP).

2 Other Assistance to Small Firms

Apart from the SIP, there are other aids available to small manufacturing firms. Some of these are general to all sizes of firms, while others are confined to small. Some are administered by public sector agencies and some by private. We now briefly outline these aids in their application to small manufacturing firms generally.

Gaeltacht Areas

We have already outlined the regional activities of the IDA and SFADCo. Small firms in Gaeltacht areas, where Irish is spoken, are catered for by Údarás na Gaeltachta established in 1979, which took over the functions of a previous body, Gaeltarra Éireann, established in 1958. The Údarás is not simply an economic development agency since it has as one of its major objectives to preserve and extend the use of the Irish language in the Gaeltacht. But it is also the body with prime responsibility for industrial development and the creation of productive employment in these areas. The agency provides advance factory accommodation, fixed asset grants (up to two-thirds of the cost), training grants, interest subsidies and after-care services. Unlike other grant-aiding bodies, the Údarás has a considerable portfolio of equity shares in companies. Total employment in all Údarás grant-aided firms amounted to just over 4,000 on 1 January 1982. This was about 500 less than the 1979 peak, but still more than double the 1973 level. In grant-aided firms with less than 50 employees, employment was 2,800 in January 1982 compared with 1,040 in 1973. According to Trident Management Consultants (1983) the total state cost per sustained job was 14 per cent higher than in the case of the IDA, not a major difference given the difficult development problems in the Gaeltacht areas. The report, however, was critical of the policy of equity participation and attempts to rescue companies during the recession by acquiring ownership. In the past few years, Údarás has concentrated on rationalising its existing portfolio of subsidiaries and associate companies and the encouragement of new industries to replace jobs lost in firms closing down.

Finance

The normal range of banking facilities is in principle open to small firms. Many small firms, however, may lack the necessary collateral or track record to meet the terms and conditions attached to bank loans. The venture capital market is still undeveloped in Ireland. Irish financial institutions in recent years have been taking steps to overcome these problems, but the scarcity of equity capital remains except for the few enterprises which have overcome the initial development problems and have prospects of further rapid growth.[34]

The Industrial Credit Company (ICC), a state-sponsored body established in 1933 to provide finance and financial services to industry, deals mainly with small and medium-sized firms. In recent years about 90 per cent of ICC projects have been with firms employing less than 100 persons. The ICC provides loans for fixed and working capital, as well as other services such as hire-purchase facilities. Although the ICC operates on a commercial basis, special facilities are available in certain circumstances. Arrangements exist between the ICC and the IDA to provide short-term loans to small industry for working capital, the loans being guaranteed by the IDA. In 1976 the ICC initiated a "Special Loans Scheme" for small under-capitalised manufacturing firms.

The ICC acts as the agent for the European Investment Bank in administering the EIB's global loan facility through which low interest loans are made available to Irish industry. These EIB loans are confined to firms employing less than 100 persons and since 1978, the Exchequer bears the exchange risk. The exact repayment period is negotiable between the ICC and the borrower within a limit of ten years. Although not more than 50 per cent of the gross cost of the project can be funded from the EIB scheme, the ICC may top up this amount through its other schemes. The ICC also administer a Venture Capital Scheme for new technology ventures where there is a good prospect of commercial viability but funds may be otherwise inadequate. Under this scheme, which has been in operation since 1978, ICC provides share capital for loans, or both, to new ventures.

One of the two main commercial banking groups, the Bank of Ireland, established an Enterprise Development Scheme in 1979 to fund small manufacturing businesses, defined as those employing less than 250 persons. This is aimed at firms considered to have good managerial talent and growth potential, but encountering difficulty in getting finance in the ordinary way. Long-term loans are provided with a moratorium on repayments in the initial years, followed by a mutually agreed repayment schedule. The two main banking groups, the Bank of Ireland and Allied Irish Banks, now have subsidiaries which are prepared to undertake equity investment in growth firms, and a number of private venture capital companies have been established in recent years. In 1981, the IDA enlisted the support of a number of institutional investors and large business firms to provide venture capital for projects under the Enterprise Development

[34]See the consultants' report in NESC (1984), referred to in Chapter 3.

Programme. Equity investment, as well as loan finance, is available in certain circumstances from An Foir Teoranta, a state-sponsored rescue company for firms in danger of bankruptcy. In 1983, funds totalling £25 million were made available to 94 firms employing about 9,000 workers.

Tax based lending arrangements have become a substantial source of finance for manufacturing firms in recent years. A variety of such arrangements were developed by financial institutions — through preference share loans, the so-called Section 84 loans, and leasing of assets — to take advantage of tax concessions to manufacturing firms (e.g., export tax relief, depreciation allowances) which would otherwise remain unused. Competition among financial institutions presumably ensured that at least part of the benefit was passed back to manufacturing firms in the form of cheap finance. None the less these arrangements have been criticised because of their arbitrary nature (see, for example, Kennedy, 1982), and the 1984 Budget significantly restricted their operation. In general, the arrangements have probably favoured large industry relatively more than small, because small industry tends to be less capital intensive and less export oriented.

The 1984 Budget also announced a tax relief scheme, resembling the UK Business Expansion Scheme, to encourage the provision of long-term risk capital for new manufacturing enterprises. Under the scheme, an investor can write off each year up to £25,000 against taxable income in respect of investment in new ordinary share issues of unquoted manufacturing and certain service companies. The scheme is subject to various conditions, including a requirement that the investment be retained for five years. The scheme has been criticised as too restrictive by Walsh (1985).

Advice and Training

Industrial training is provided for industry generally by AnCO which operates in close liaison with the IDA in the case of SIP firms. The Irish Management Institute (IMI) in conjunction with the IDA operates a Business Development Programme for managers of small firms. The programme aims to help managers identify opportunities for expansion and to plan future growth. A Business Advisory Service for firms employing less than 200 persons is provided by the Irish Productivity Centre (IPC), an autonomous agency established jointly by the central employer and union organisations and receiving a state grant. The typical firm dealt with by the IPC is a family business, employing 15-70 persons. A subsidised fee is charged for each assignment. In 1983, 148 assignments were completed. The IPC may also act as an adviser to improve industrial relations in a situation where an industrial dispute is imminent.

The Institute for Industrial Research and Standards (IIRS), with an annual government grant-in-aid in 1985 of £8 million, provides a variety of technical advisory services, generally on a subsidised fee basis. The Kilkenny Design Workshop gives advice on design of products as well as on the selection of suitable materials and on production techniques: the turnover on its design and training services is in excess of £1 million. As regards marketing, Coras Trachtala (CTT, the Irish Export Board) has recently established a department to cater specifically for small-exporting firms. CTT provides grants towards the cost of undertaking market research in overseas market areas, as well as grants towards identifying new products. Its grant-in-aid for all activities is £23½ million in 1985. The Irish Goods Council, with a grant-in-aid of £1 million in 1985, advises firms on the marketing of products on the home market, and promotes subcontracting. The IIRS and NBST co-operate in a scheme for the placement of engineering graduates in small firms, funded by the Youth Employment Agency. A similar scheme is in operation for the placement of marketing graduates in small firms. The NBST also acts as agent to facilitate R & D projects in colleges for firms, the funding being provided by the IDA. In 1984, projects amounting to about £2 million were initiated in this way.

Taxation

Tax arrangements for manufacturing firms are much the same for all sizes of firms. There is free depreciation of fixed assets, which allows firms to write-off up to 100 per cent of the value in one year. The amount that can be written off is reduced by the value of IDA grants in the case of industrial buildings but not in the case of plant and machinery. Up to 1981, profits on export sales were exempt from tax. This scheme is now being phased out, and replaced by a corporation tax of 10 per cent on manufacturing profits, whether the output is sold at home or abroad, compared with the standard rate of 45 per cent. Manufacturing companies with total taxable profits of less than £25,000 in one year are entitled to relief which reduces the effective tax rate from 10 per cent to 7.8 per cent. Marginal relief is allowed on taxable profits between £25,000 and £35,000.

3 *Conclusions*

Programmes for developing small manufacturing industry have expanded considerably over the past 15 years or so. They involve, *inter alia*, a sizeable commitment of public funds. There has been little research, however, into the impact of these programmes on the ground, or their effectiveness compared to other programmes. In the next two chapters of this Part, we try to go some way to filling that gap in regard to the SIP, but there is need for research also into many of the other programmes described briefly above.

There is always the danger that the incremental growth of programmes could lead to an undue proliferation of services and agencies, since rarely are existing measures dropped when new ones are added. Moreover, with the passage of time it becomes necessary to enquire whether the original rationale of the general strategy still holds. We will take up these questions in Part III.

Chapter 8

EMPLOYMENT IN SIP FIRMS

This chapter first examines the level and trend of employment in firms assisted under the Small Industry Programme (SIP) compared with other sectors of manufacturing industry from 1973 to 1980. The second section then examines the structure of employment in SIP firms in terms of industry group, region etc.

1 The Level and Trend of SIP Employment

According to the IDA *Annual Employment Survey* for 1980 there was on 1 January 1980 a total of 1,468 establishments which had been approved up to then for a SIP grant (including those handled by SFADCo), and their total employment was 24,246 (Table 8.1). It must be strongly emphasised, however, that although we refer to this as SIP employment, it does not mean that the SIP was responsible for the creation of that many jobs. Many of the recipients of SIP assistance are established firms, and the figure given above is the total employment of all SIP recipients, including that which existed before they were approved for SIP grants. While, as mentioned in the previous chapter, firms may begin business or expand in anticipation of approval under the SIP, nevertheless a closer approximation to the amount that might be properly credited to the effect of the SIP would be got by deducting employment already existing at the time of first approval.

Even such a figure, however, would only be a first approximation for the following reasons. First, some of the employment growth could have taken place anyway even if there had been no SIP. By the same token, there could have been employment decline in some firms had they not received SIP assistance. It would be difficult, if not impossible, to assess where the balance lies between the two opposing factors. Second, since most small firms serve the home market, there is a possibility that increased employment in SIP firms may displace employment in existing firms. The IDA seek to minimise this in the administration of the programme: as mentioned in the previous chapter grants are not normally made to activities where there is overcapacity or a shortage of raw materials. Neverthe-

Table 8.1: *Employment in SIP establishments 1973 and 1980**

	1973			*1980*		
	No. of Establishments	*Employment*	*Average size*	*No. of Establishments*	*Employment*	*Average size*
Began business after 1973	—	—	—	728	9,658	13.3
In business in 1973	907	12,000	13.2	740	14,588	19.7
Total	907	12,000	13.2	1,468	24,246	16.5
Breakdown of "in-business" in 1973:						
Gone out of business by 1980	167	1,880	11.3	—	—	—
Experienced employment decline	216	3,592	16.6	216	2,411	11.2
Experienced employment increase or no change	524	6,528	12.5	524	12,177	23.2
Sub-total	907	12,000	13.2	740	14,588	19.7

*The data show the number of establishments and their employment in 1973 and 1980 for all establishments which received approval for a SIP grant at any time from 1967 to 1979, and which were in business in either 1973 or 1980. They do not therefore include (i) establishments which were approved for a SIP grant and which had gone out of business before 1973; (ii) establishments which had come and gone in the years 1974–79; or (iii) establishments which had not started by 1 January 1980.

Source: IDA *Annual Employment Survey 1980*. In this and all subsequent tables, SFADCo firms are included unless specifically excluded.

less, it is possible that some displacement happens. Perhaps it is not altogether undesirable if the new firms are much more dynamic than the existing ones. Be that as it may, it still reduces the net gain in employment attributable to the SIP. It is particularly difficult, however, to assess lost production and employment due to the failure, shrinkage or retarded growth of firms affected indirectly by favouring their competitors. Third, the SIP category includes those firms which were expanding to such a degree that they later qualified for a NIP grant. In 1980 employment in establishments approved up to then for both SIP and NIP grants amounted to 4,159, compared with 1,985 in the same establishments in 1973.

Unfortunately the data available do not make it possible to quantify these factors in the way that we would wish. The IDA Employment Survey was first conducted in 1973 and no satisfactory data are available on actual employment in

SIP firms before then. We show in Table 8.1 the actual employment in 1973 and 1980 in different categories of firms which had been at any time up to end-1979 approved for a SIP grant. Of the total of 1,468 such etablishments in 1980, 728 were not in existence at all in 1973 and it may be taken that all of these were at least influenced by the prospect of SIP assistance, even if some were set up before being approved for a grant. If we aggregate the employment in these new entrants in the year of first recorded employment, it amounted to 5,961, and had risen to 9,658 in 1980.

In regard to the establishments already in business in 1973, it is not possible to give any indication of how much of the net increase of 2,588 in employment took place after they had received approval for a SIP grant: while some were approved after 1973 when their employment would in most cases have been higher than in 1973, others were approved prior to 1973 when their employment in most cases would have been lower. What can be said is that 18 per cent of all such establishments had gone out of business between 1973 and 1980, 24 per cent had experienced a fall in employment, while a majority, 58 per cent, had increased or maintained their 1973 level of employment.

The average size of the SIP establishments rose from 13.2 persons in 1973 to 16.5 persons in 1980. This rise was mainly due to the growth of establishments already in existence in 1973, since average size in 1980 of the new entrants during the period, 13.3, was much the same as the overall average in 1973. The firms which went out of business after 1973 had an average size in that year which was below the overall average.

Comparison with other Categories of Industry

Table 8.2 compares experience in SIP establishments with other categories of industry. The reader is reminded that, for the reasons outlined above, what we are portraying is the comparative experience of those firms approved for an IDA grant, rather than ascribing that experience to the effect of the SIP. The table distinguishes four other categories as follows. "NIP/indigenous" refers to establishments approved for a new industry grant at any time up to and including 1979 and whose main corporate headquarters is situated in the Republic of Ireland. "NIP/foreign" comprises establishments approved for a new industry grant at any time up to and including 1979, and whose main headquarters is located outside the state. As mentioned, some establishments have been approved for both NIP and SIP grants, and these are included in the SIP category and excluded from the NIP category. All establishments which did not receive a NIP and SIP grant are included under "Other large" and "Other small" categories, depending on whether they employed more or less than 50 persons in 1973, or in the year of first employment after 1973 in the case of new entrants. Most of these establishments are indigenous and some would have been

Table 8.2: *Components of employment changes in various types of IDA grant-aided and other establishments, 1973–1980*

				Gross changes					
				Gains			*Losses*		
	1973 total employment	*1980 total employment*	*Total net change 1973–80*	*New entrants*	*Employment increases*	*Total gains*	*Closures*	*Employment decreases*	*Total losses*
SIP Industry	12,000	24,246	12,246	9,658	5,649	15,307	1,880	1,181	3,061
NIP: indigenous	49,624	50,820	1,196	5,115	9,258	14,373	6,500	6,677	13,177
NIP: foreign	35,566	58,954	23,388	24,567	9,259	33,826	5,593	4,845	10,438
Other Small	31,711	33,683	1,972	6,825	6,382	13,207	7,538	3,697	11,235
Other Large	88,858	74,815	-14,043	5,625	6,544	12,169	11,529	14,683	26,212
Total	217,759	242,518	24,759	51,790	37,092	88,882	33,040	31,083	64,123

Note:
"Other Small" represents establishments which employed between one and 49 persons in 1973 (or in the first year of entry subsequently) and which did not receive a SIP or NIP grant.
"Other Large" represents establishments with employment of 50 or more persons in 1973 (or in the first year of entry subsequently) and which did not receive a SIP or a NIP grant.
"New entrants" are those establishments which had no employment in 1973 but had in 1980. New entrants which did not receive a NIP or a SIP grant were included in "Other Small" where they began with less than 50 persons engaged; otherwise they were included in "Other Large".
"Increases" refer to employment changes in establishments in existence in both 1973 and 1980 and where the 1980 employment was greater than or equal to the 1973 level.
"Closures" refer to job losses in establishments which were in employment in 1973 but had closed by 1980.
"Decreases" refer to employment changes in establishments in existence in both 1973 and 1980 and with lower employment in the latter year.
Source: IDA *Annual Employment Survey 1980*. The qualifications to the data noted in Chapter 2 (pp. 21 and 24) and in Table 4.6 are relevant here also.

approved for a grant under other IDA programmes offering re-equipment, research and development and training grants.

As may be seen from Table 8.2 by far the largest contribution to the increase in total manufacturing employment was in foreign NIP establishments, with a net increase of almost 24,000. But the rise of 12,000 in SIP establishments was also considerable. In fact, as may be seen from Table 8.3, which shows the percentage increases on the 1973 level, the relative rise in SIP establishments, 102 per cent, was higher than in the foreign NIP sector, 66 per cent. The employment performance of domestic industries with 50 or more employees has been dismal. Among those assisted by NIP grants, there was a net increase of only two per cent, while there was a decline of 16 per cent in "other large" establishments (mainly domestic). Though small establishments, other than those assisted under the SIP, had only a small net gain in employment, nevertheless, their employment performance was better than in large indigenous industry, whether grant-aided or not.[35]

It is clear from Table 8.3 that grant-aided establishments, whether large or small, did better than their counterparts which were not grant-aided. In evaluating this, however, it must be remembered that SIP and NIP grants are on the whole given to firms with at least some growth prospects; while firms which are likely to decline anyway do not normally receive such assistance. Hence the superior performance partly reflects the success of the IDA in picking those firms most likely to grow anyway, and cannot therefore be entirely attributed to the benefit of the grant. Even recognising this point, however, it, nevertheless, seems that among indigenous firms, the SIP sector (which is overwhelmingly indigenous) performed best in this period in terms of employment. As a proportion of total manufacturing employment, SIP establishments had almost doubled their share from 5.5 per cent in 1973 to 10 per cent in 1980.

Tables 8.2 and 8.3 also decompose the net employment change into (a) total gross gains (the aggregate of employment increases in establishments recording a net increase over 1973-80) and (b) total gross losses (the aggregate of employment decreases in establishments recording a net decrease over 1973-80). The difference between total gross gains and total gross losses gives the total net change in employment. Total gross gains are further decomposed into gains attributable to (i) new entrant establishments (i.e., those commencing employment later than 1973 and which survived to 1980), and (ii) employment expansion in establishments which were in employment in both 1973 and 1980. Total gross losses are also divided into two categories: (i) establishments in employ-

[35]As mentioned in Chapter 2, O'Farrell (1984a) has suggested that the IDA Survey data understate the extent of closures in small establishments in Dublin. The conclusions above, however, remain true when Dublin is excluded from the data.

Table 8.3: *Percentage change in employment in various types of industry (1973–80)*[a]

		Gains			Losses		
	Total net change in employment 1973–1980	*New entrants*	*Employment increases*	*Total gains*	*Closures*	*Employment decreases*	*Total losses*
	%	%	%	%	%	%	%
SIP Industry	102.1	80.5	47.1	127.6	15.7	9.8	25.5
NIP: indigenous	2.4	10.3	18.7	29.0	13.1	13.5	26.6
NIP: foreign	65.8	69.1	26.0	95.1	15.7	13.6	29.3
Other Small	6.2	21.5	20.1	41.6	23.8	11.7	35.4
Other Large	-15.8	6.3	7.4	13.7	13.0	16.5	29.5
Total	11.4	23.8	17.0	40.8	15.2	14.3	29.4

[a]Percentage changes in employment are in all cases expressed relative to total employment in each category in 1973.
Source: As in Table 8.2.

ment in 1973 which closed down before 1980, and (ii) establishments which were in employment in both 1973 and 1980 and recorded a decline in employment. In comparing the position between 1973 and 1980, we are seeking to exclude as far as possible purely short-term fluctuations. Thus establishments which commenced employment later than 1973 and subsequently ceased before 1980 are not included since they do not affect the net change for 1973–80.

Looking first at employment losses, it will be seen from Table 8.3 that the total losses in each of the different categories of industry, as a percentage of total 1973 employment in each category, did not vary very much from the overall figure of 29½ per cent. The highest rate of losses was in "other small" establishments (35½ per cent) and lowest in SIP establishments (25½ per cent). The pattern of losses as between closures and employment decreases did vary somewhat, however, with the two categories of small industry experiencing relatively more losses through closures rather than employment decline, as compared with larger firms. This pattern, as pointed out in Part I, is quite general in other countries also.

Given that the rate of employment loss did not vary greatly among the different categories of industry, it is only to be expected that the large differences in net employment changes reflect differences in gross gains. This in fact was the position, with the SIP and foreign NIP sectors experiencing substantially higher rates of gross employment gains. It is of interest to note, however, that in both cases the greater part of those gains was due to new entrants. Indeed if new entrants are excluded, no category except the SIP sector had a sufficiently high rate of employment increase to offset its rate of employment losses. Those categories, therefore, which had few new entrants had the worst overall net employment experience.

That the entry rate, apart from the SIP sector, was low for indigenous industry, whether grant-aided or not, is not perhaps surprising in the case of wholly new enterprises — as distinct from a new establishment owned by an older firm. Such enterprises are most likely to start small, and the more promising ones would be likely to qualify for SIP assistance. What is disappointing is the poor record of established large indigenous industry which has neither shown much expansion in existing establishments, nor any strong tendency to diversify into new activities in separate large new establishments. Without the contribution of small industry, and especially the SIP section of it, the performance of indigenous industry would have been very poor indeed.

2 *Structure of Employment in SIP Industry*

Industry Group

Table 8.4 gives a breakdown of SIP establishments and employment by industry group in 1973 and 1980. The dominant group is metals and engineering,

Table 8.4: *SIP establishments and employment distributed by industry group 1973 and 1980*

Industry group	*1973*		*1980*		
	No. of establishments	*Employment*	*Number of establishments*	*Employment*	*% change in employment 1973–1980*
1 Food	55	940	79	1,553	65.2
2 Drink and tobacco	4	35	4	84	140.0
3 Textiles	63	991	74	1,168	17.9
4 Clothing and footwear	51	947	90	1,880	98.5
5 Wood and furniture	149	2,282	196	3,053	33.8
6 Paper and printing	31	490	49	838	71.0
7 Chemicals	18	236	41	680	188.1
8 Minerals	81	876	122	1,491	70.2
9 Metals and engineering	353	4,039	614	10,721	165.4
10 Miscellaneous manufacturing	102	1,164	199	2,778	138.7
Total	907	12,000	1,468	24,246	102.1

Source: IDA *Annual Employment Survey 1980*.

which accounted for over 40 per cent of all employment in SIP establishments in 1980 and was also one of the groups that increased most from 1973–80. This is not surprising given that small establishments are especially suited to sub-supply business in providing components, tools and specialised machining operations for use in medium-to-large-scale firms. Demand for such work increased rapidly in the 1970s with the high rate of investment in plant and machinery — not only in manufacturing but also in agriculture and the construction industry. As already mentioned in Chapter 3, much of the sub-supply, however, has been in low-skill areas, and only a small, though rising proportion, of the sub-supply needs of the large new foreign companies is met domestically (O'Farrell and O'Loughlin, 1980; 1981 and O'Farrell 1982).

The next most important group was wood and furniture, accounting for one-eighth of total employment in 1983. In most countries this group is one in which small-scale establishments account for a relatively high proportion of total employment in the industry. The growth of SIP employment in this group was relatively low, however, reflecting the high degree of competition from imports. Only one other group, miscellaneous manufacturing, accounted for more than 10 per cent of SIP employment in 1980. The three groups mentioned — metals and engineering, wood and furniture and miscellaneous manufacturing — together accounted for over two-thirds of total SIP employment in 1980.

These three groups are also the only ones in which the share of SIP employment in relation to total employment in the group was above average in 1980. This may be seen from Table 8.5, which shows the ratio of SIP employment in each group in 1980 to various other aggregates in the same group. For manufacturing as a whole the share of SIP employment in total employment was 10 per cent in 1980, but in wood and furniture it amounted to 27 per cent, in metals and engineering to 18.6 per cent and in miscellaneous manufacturing to 11.9 per cent. Industry groups with a particularly low proportion were drink and tobacco (0.7 per cent), food (3.2 per cent), chemicals (5 per cent), and paper and printing (4.8 per cent). Drink and tobacco and chemicals are highly capital intensive activities which typically have few small-sized establishments. Furthermore, the IDA have been reluctant to aid establishments in the soft drinks and bottling industries because of existing excess capacity, a factor which applies also in the printing industry. Perhaps most surprising is the fact that there are so few SIP establishments in the food sector. Unless, however, new firms were producing a novel product or were substantially involved in export, they would be in direct competition with existing producers, while the marketing expense required to launch a new food product or to enter export food markets would generally require resources far beyond those available to most small firms. The Telesis Report (1982) estimated that just over one-half of IDA grant payments to indegenous SIP firms in the period 1967–79 were in respect of non-traded activities, defined as those where the key to competitive success lies in transport costs or logistic factors.

Table 8.5 also attempts to give a picture of the relative importance of SIP employment in each group relative to employment in all small establishments in the group. In order to compare like with like, the figures shown in Col. 2 of that table represent the ratio of employment in SIP establishments with less than 50 persons in 1980 to employment in all establishments with less than 50 persons. Hence the numerator does not include the full total of SIP employment in the group, since it excludes those SIP establishments that have grown to 50 or more persons. For manufacturing as a whole, it emerges that nearly one-third of total employment in small industry is in establishments that have been approved for a small industry grant. The proportion is much higher in metals and engineering (47.9 per cent) and textiles (40.8 per cent). Drink and tobacco (6.9 per cent), food (10.1 per cent) and paper and printing (13.7 per cent), again all emerge with the lowest proportions — a reflection of the point made earlier about excess capacity already in those industries for the general run of such products sold on the domestic market.

Finally, Column 3 of Table 8.5 shows the ratio of SIP employment to total employment in IDA grant-aided firms in each group in 1980. "Grant-aided" here

Table 8.5: *SIP employment as a proportion of total employment by industry group, 1980*

	SIP as a % of		
Industry group	*Total*	*Total <50*	*Total IDA grant-aided*[a]
1 Food	3.2	10.1	6.6
2 Drink and tobacco	0.7	6.9	5.3
3 Textiles	5.8	40.8	7.9
4 Clothing and footwear	9.0	27.7	19.0
5 Wood and furniture	27.0	35.3	62.9
6 Paper and printing	4.8	13.7	21.6
7 Chemicals	5.0	28.2	7.7
8 Minerals	8.5	32.8	17.4
9 Metals and engineering	18.6	47.9	25.6
10 Miscellaneous manufacturing	11.9	37.3	17.3
Total	10.0	31.1	18.1

[a]Relates to employment in establishments which received SIP or NIP grants.
Source: IDA *Annual Employment Survey, 1980.*

refers to NIP and SIP grants, whether given to foreign or indigenous industry. Overall, 18 per cent of employment in all grant-aided establishments in 1980 was in the SIP sector. The groups with the largest SIP shares were wood and furniture (62.9 per cent) and metals and engineering (25.6 per cent). The high figure for wood and furniture reflects not only the predominance of small firms but also the fact that there have been relatively few NIP grants in this sector, and that their performance in terms of employment has been relatively poor. Similar factors account for the high SIP share in paper and printing.

Table 8.6 shows the proportion of female employment in SIP industry and in total manufacturing in 1980, classified by industry group. Overall, there was a slightly lower share of female employment, 24.4 per cent in SIP industry than in total manufacturing, 29.2 per cent. But this arises largely due to the different composition of industry in the two aggregates. As may be seen from Table 8.6, in every group the female share was higher in the SIP sector, with the exception of metals and engineering where SIP firms had a considerably lower female share. But, as already pointed out, there is an exceptionally high concentration of SIP employment in the metals and engineering and wood and furniture groups, and these are groups with a low female share.

Region

As may be seen from Table 8.7, which gives the regional breakdown, SIP em-

Table 8.6: *Female employment share in SIP sector and in total manufacturing 1980*

Industry group	*SIP*	*Total manufacturing*
	%	%
1 Food	33.6	24.2
2 Drink and tobacco	22.6	20.8
3 Textiles	68.9	39.9
4 Clothing and footwear	77.0	74.3
5 Wood and furniture	11.0	10.3
6 Paper and Printing	32.7	29.2
7 Chemicals	38.4	30.7
8 Minerals	15.5	13.1
9 Metals and engineering	9.9	22.4
10 Miscellaneous manufacturing	34.9	31.4
Total	24.4	29.2

Source: IDA *Annual Employment Survey, 1980.*

Table 8.7: *SIP establishments and employment distributed by region 1973 and 1980*

	1973			*1980*			
			SIP employment			*SIP employment*	
Region	*Number of establishments*	*Employment*	*Total manufacturing employment %*	*Number of establishments*	*Employment*	*Total manufacturing employment %*	*% change in employment 1973–1980*
North-East	96	1,590	8.9	127	2,868	15.6	80.4
South-East	115	1,469	6.5	185	3,188	11.1	117.0
Midlands	103	1,215	13.4	139	2,265	15.6	86.4
South-West	115	1,536	4.5	188	3,492	9.5	127.3
Mid-West	81	1,197	6.6	161	2,216	11.1	85.1
West	93	779	9.3	124	1,505	9.7	93.2
North-West	100	1,385	16.0	135	1,984	16.8	43.2
East (excl. Dublin)	74	976	6.5	134	2,257	11.9	131.3
Total (excl. Dublin)	(777)	(10,147)	(7.6)	(1,193)	(19,775)	(12.0)	(94.9)
Dublin	130	1,853	2.2	275	4,471	5.8	141.3
Total	907	12,000	5.5	1,468	24,246	10.0	102.1

Source: IDA *Annual Employment Survey, 1980.*

ployment is fairly evenly spread across the nine planning regions. In 1980, the highest share was in Dublin with 18.4 per cent, and the lowest in the west, with 6.2 per cent. As a proportion of total manufacturing employment in the region, however, Dublin's share at 5.8 per cent was the lowest, while the north-west with 16.8 per cent was the highest.

In every region, the growth of SIP employment over the period 1973–80 was greater than in total manufacturing in the same region — as is indicated by the fact that the share rose in all cases. The highest rates of growth in SIP employment were experienced in Dublin (141.3 per cent) and the east, excluding Dublin (131.3 per cent), while the lowest rate was in the north-west (43.2 per cent).

It would appear that the inter-regional differences in growth are largely related to the extension of the range of the SIP programme over time. As mentioned in Chapter 7, the programme started in the designated areas in 1967, was extended in 1969 to all areas except Dublin, and was applied to Dublin for some products in 1975 and for nearly all products in 1977. This mainly accounts for the tendency for those regions with the lowest relative amount of SIP employment in 1973 to have the fastest growth in SIP employment from 1973 to 1980: if we correlate the ratio of SIP employment to total manufacturing employment in each region in 1973 with the growth of SIP employment in the region from 1973 to 1980, the correlation coefficient is -0.88, which is statistically significant at the 0.01 per cent level. This point is reinforced by the fact that in Dublin, out of total employment in all establishments with less than 50 persons in 1980, only 20.7 per cent was in establishments approved for a SIP grant, as against 35.6 per cent in the rest of the country and 45.9 per cent in the north west.

Nationality

It is clear from Table 8.8 that the SIP establishments are overwhelmingly indigenous in origin. The nationality of an establishment is defined by the location of its main corporate headquarters. Irish establishments constituted 95.8 per cent of the total in 1973 and 92.1 per cent in 1980. The average size of establishment was only slightly higher in foreign than in Irish establishments, so that the Irish shares in total SIP employment were very similar to the establishment shares. Employment in foreign establishments increased four-fold from 1973 to 1980, as against an approximate doubling of employment in Irish SIP establishments, reflecting a relatively larger rate of new entry among foreign establishments. UK establishments accounted for 40 per cent of employment in foreign SIP establishments in 1973. Growth from 1973 to 1980 was faster, however, in non-UK establishments and in 1980 the amount of employment in UK SIP establishments (433) was less than in German (535) or US (462) establishments.

Table 8.8: *SIP establishments and employment by nationality, 1973 and 1980.*

Country	*1973*			*1980*			
	No. of Establishments	*Employment*	*Average size of establishment*	*No. of establishments*	*Employment*	*Average size of establishment*	*% change in employment*
Ireland	869	11,481	13.2	1,352	22,131	16.4	92.8
United Kingdom	18	203	11.3	39	433	11.1	113.3
Germany	8	121	15.1	24	535	22.3	342.1
United States	8	131	16.4	21	462	22.0	252.7
Other European	3	39	13.0	29	589	20.3	1,410.3
Other non-European	1	25	25.0	3	96	32.0	284.0
Sub-total: Foreign	(38)	(519)	(13.7)	(116)	(2,115)	(18.2)	(307.5)
Total: All Nationalities	907	12,000	13.2	1,468	24,246	16.5	102.1

Source: IDA *Annual Employment Survey, 1980.*

Size

Table 8.9 shows the size distribution of SIP establishments and their employment in 1973 and 1980. Of the total of 1,428 SIP establishments in 1980, more than 70 per cent (1,061) had less than 20 persons engaged, but they accounted for slightly less than one-third of the employment (7,925 out of 24,246). These very small establishments, however, have shown a considerable capacity for growth, which cannot be seen by looking simply at the size distribution in different years, changes in which are the outcome of firms moving up or down in size class, and into or out of business. Table 8.9, however, also presents the change in employment from 1973 to 1980 keeping the 1973 size classification constant. These figures show the employment performance of the establishments in a given size classification in 1973. The net employment change from 1973 to 1980 in the establishments with less than 20 persons in 1973 was 1,610 (or 32 per cent) compared wth 978 (or 14 per cent) in the establishments with 20 or more persons. As regards new entrants, also shown in Table 8.9, establishments with less than 20 persons (in 1980) accounted for 80 per cent of the new SIP establishments and 45 per cent of employment.

Particular interest centres on the ability of small firms to develop into larger ones. It may be seen from Table 8.9 that in 1973 there were 21 SIP establishments with 50 or more persons, having a total employment of 1,389 (11½ per cent of the total), while in 1980 there were 76 such establishments employing 6,373 (or 26 per cent of the total). Of the 21 establishments with 50 or more employees in 1973, 10 had increased their employment, nine had suffered declines, one had experienced no change and one had closed down by 1980. The overall net change in their employment from 1973 to 1980 was a rise of 411 or 23 per cent. Of the 76 SIP establishments with 50 or more persons engaged in 1980, 13 had 50 or more in 1973, 49 expanded from less than 50 in 1973 and 14 were new entrants. In the case of the 14 new entrants their employment in 1980 was 1,262 compared with 549 in the year they entered, while for the 49 which existed in 1973 the overall increase was from 1,408 in 1973 to 3,554 in 1980.

Table 8.10 gives a picture of the degree to which the 76 establishments in 1980, with more than 50 persons, expanded over the period. In regard to those in existence in 1973, 22 of the 62 establishments had moved up from less than 30 in 1973, but of those only one had more than 100 workers in 1980. Of the 27 establishments with 30-49 workers in 1973, six had over 100 in 1980. Only one establishments had more than 200 persons in 1980 and this was already in the size range 50-99 in 1973. Among the 14 new entrants which had more than 50 persons by 1980, two of them had over 200 persons, having started in their first year in the range 30-49. While the figures suggest a capacity on the part of a small minority of firms to move from small to medium size, the numbers involved are few as yet, and there are no really dramatic cases, like those in the US, of new

Table 8.9: *SIP establishments and employment by size, 1973 and 1980*

Size classification	*1973* *No. of establishments*	*1973* *Employment*	*1980* *No. of establishments*	*1980* *Employment*	*Employment change 1973–80 in 1973 establishments*	*New entrants (1980 size classification)* *No. of establishments*	*New entrants (1980 size classification)* *Employment*
1– 4	285	714	388	924	493	203	494
5– 9	199	1,361	325	2,224	834	180	1,231
10– 14	135	1,593	220	2,645	347	122	1,468
15– 19	80	1,335	128	2,132	-64	68	1,131
20– 29	101	2,412	182	4,342	6	86	2,069
30– 49	86	3,196	149	5,606	561	55	2,043
50– 99	20	1,265	61	3,856	363	11	705
100–199	1	124	12	1,773	48	1	138
200–499	—	—	3	744	—	2	419
Total	907	12,000	1,468	24,246	2,588	728	9,658

Source: IDA *Annual Employment Survey, 1980*

Table 8.10: *Size class in 1973 (or first year of entry) of establishments with 50 or more persons in 1980*

	Size class 1980			
	50–99	*100–199*	*200–499*	*Total*
Size Class 1973	*No. of Establishments in Existence in 1973*			
<20	12	1	—	13
20– 29	9	—	—	9
30– 49	21	6	—	27
50– 99	8	3	1	12
100–199	—	1	—	1
Total	50	11	1	62
Size Class: Year of Entry	*New Entrants*			
<20	2	1	—	3
20–29	1	—	—	1
30–49	5	—	2	7
50–99	3	—	—	3
Total	11	1	2	14

Source: IDA *Annual Employment Survey, 1980.*

small firms quickly moving into the large category of 500 or more employees.[36]

3 *Conclusions*

Employment in establishments approved for SIP grants at any time up to end-1979 rose from five and a half per cent of total manufacturing employment in 1973 to 10 per cent in 1980. Not all of this can be attributed to the effects of the SIP alone, no more than the growth in NIP industry can be attributed solely to the new industry grants. The figures show, however, that the performance in small firms in general, and SIP firms in particular, made a significant contribution to the overall employment performance in this period.

SIP establishments are overwhelmingly indigenous and are heavily concentrated in the metals and engineering and the wood and furniture industrial groups. The employment is widely dispersed regionally, confirming the a priori expectation that small industry can be particularly adaptable to locations that could not sustain a large industry. Some SIP firms have shown a capacity to grow into larger firms but they are as yet a very small minority.

[36]Similar findings are reported in O'Farrell (1984b,c, and 1985a).

The heavy dependence of many SIP firms on local domestic demand obviously raises a question about their potential for expansion, unless more can be encouraged to engage directly or indirectly in exports. This is particularly so given that the agricultural boom of the 1970s is unlikely to be repeated, and that construction activity is severely constrained by the state of the public finances. On the other hand, the fact that only a small proportion of the sub-supply needs of new large manufacturing firms were met domestically suggests that there may be considerable scope for developing linkages provided the skill levels of sub-supply small firms can be upgraded. These issues will be taken up again in the final policy chapter.

Chapter 9

SIP GRANTS AND INVESTMENT

In this chapter we present data on grants received and investment undertaken by SIP firms. We consider first the data on commitments made at the time projects were approved, and we then examine the rate at which intentions were translated into actuality. Finally, we look at the record of failures as measured by firms closing down. In evaluating the results, comparisons are made with other IDA programmes.

1 *Approvals Data*

Decisions to award industrial grants are normally taken on the basis of planned projects rather than projects actually undertaken. The data relating to these decisions, commonly known as approvals data, refer therefore to future intentions or commitments rather than actual achievements. For a variety of reasons, the actual experience differs from that envisaged at the approval stage. Projects may be abandoned or delayed due to a change in market circumstances, or in the ability of the promoter to proceed. They may be scaled up or down in size, or altered in other ways. Clearly the actual outcome is what is most important to the economy. Nevertheless, the approvals data are of some interest in themselves in giving an indication of the planning experience of the development authorities. The approvals represent commitments, subject to certain conditions, on behalf of the development authorities, and it is useful to examine the nature of these commitments, before going on to consider the rate and timing of their translation into actuality.[37]

[37]A large amount of data is published each year in the IDA *Annual Report*, and the associated *Small Firms Annual Report* first published in 1977 and then entitled *Small Industries Programme: Annual Report 1977 and Review 1967–77*. We also had access to the IDA Project Information File, which contains the detailed records for each project. In regard to the mid-west region, where the SIP has been administered by SFADCo since 1978, information was kindly supplied by SFADCo.

Number of SIP Projects

Table 9.1 gives details of the number of SIP projects approved each year from 1967 to 1982. The numbers grew rapidly in the early years as the programme got under way and was extended to wider areas. In the first half of the 1970s there was a downward trend in the number of projects approved, influenced partly by the general recession in the years 1974–75. The renewed increase thereafter derived from a number of factors: the more buoyant economic conditions, greater aggressiveness in promoting the SIP, and the inclusion of Dublin in the scope of

Table 9.1: *SIP grant approvals 1967–1982*[a]

Year	*No. of Projects*	*Total grants approved*[b] *(current prices) £'000*	*Total grants approved (constant 1982 prices*[c]*) £'000*	*Average grant per project (constant 1982 prices) £'000*
1967	22	65	378	17.2
1968	70	718	4,064	58.1
1969	248	2,040	10,923	44.0
1970	220	1,313	6,372	29.0
1971	135	891	3,997	29.6
1972	136	1,283	5,319	39.1
1973	148	1,845	6,939	46.9
1974	144	2,844	8,694	60.4
1975	129	2,302	5,872	45.5
1976	193	5,489	12,135	62.9
1977	216	6,025	11,145	51.6
1978	469	12,451	20,760	44.3
1979	735	25,194	36,642	49.9
1980	994	36,313	45,077	45.3
1981	890	32,259	35,334	39.7
1982	818	29,748	29,748	36.4
Total	5,567	160,780	243,399	43.7

[a]The data here relate to manufacturing projects, and exclude services activities. Also excluded are projects under the Enterprise Development Programme.

[b]Includes grants for training and rental of factory space as well as grants for acquisition of fixed assets. Grants approved by SFADCo. are included.

[c]For derivation of the constant price data, see Appendix 1.

Sources: IDA *Project Information File, Small Firms Annual Report* and Informatin supplied by SFADCo.

the SIP. A peak of 994 approvals was reached in 1980, and the sharp decline thereafter must again be attributed to the general recession in economic activity.

It should be explained that the number of projects approved is not synonymous with the number of establishments which received SIP approval. Grant approvals may be staged over a number of years to allow time to evaluate the impact of initial payments, or in conformity with a phased programme of development for the firm. An analysis of the data in the Project Information File for the years 1967–78 revealed that up to 1978 about 80 per cent of establishments had been approved only once, a further 14 per cent had received two project approvals and the remaining six per cent of establishments had received three or more approvals. The aggregate of repeat approvals from 1967–78 amounted to 22 per cent of all approvals in the period.

Table 9.2 gives the industrial and regional breakdown of the number of projects approved since 1967. As might be expected from the information in the previous chapter on SIP employment, the metals and engineering group dominates, accounting for 37 per cent of all project approvals in the period 1967–82. Two other groups between them, wood and furniture and miscellaneous manufacturing, account for a further 32 pr cent of the total. Over time the major sectoral change was a considerable rise in the share of miscellaneous manufacturing from 10 per cent of the total in the five years 1967–71 to 17 per cent in the years 1977–81, matched by a similar decline in textiles from 10½ to 3½ per cent.

In regard to the regions, the pattern has tended to follow the extension of the SIP to different areas over time. Thus while the east region (including Dublin) accounted for just under 10 per cent of projects in the first five years, it now accounts for over one-third. Much of the growth in the total number of SIP project approvals has come about in the regions where the programme was least important in the early years. On the other hand, in the north-west which accounted for 18 per cent of approvals in the first five years, the increase was lowest and the region accounted for only 6 per cent in the five years 1977–81.

Grants Approved

As may be seen from Table 9.1, the total grants approved for SIP projects in the period 1967–82 amounted to £243m. at constant 1982 prices. The average real grant per project has fluctuated a good deal but has not shown any upward trend. Thus the rising trend in the aggregate amount of grants approved has been due to the increased number of projects. In the early years the grants were almost entirely capital grants for the acquisition of fixed assets but with the passage of time other grants — in respect of training, rent reduction and leasing of equipment — have become more important and accounted in 1982 for nearly a quarter of total expenditure.

Table 9.3 gives the industrial and regional breakdown of grants approved

Table 9.2: *Industrial and regional distribution of number of SIP project approvals, various sub-periods, 1967–1982*

	1967–71	*1972–76*	*1977–81*	*1982*	*Total 1967–82*
Industry Group					
Food	44	54	164	57	319
Drink & tobacco	3	1	10	1	15
Textiles	73	41	115	29	258
Clothing & footwear	43	42	269	70	424
Wood & furniture	112	92	510	198	912
Paper & printing	19	20	131	37	207
Chemicals	6	25	87	24	142
Minerals	66	68	166	47	347
Metals & engineering	259	284	1,296	323	2,071
Miscellaneous	70	123	556	123	872
Total	695	750	3,304	818	5,567
Region					
North-east	79	95	228	53	455
South-east	99	92	308	68	567
Midlands	92	87	211	46	436
South-west	92	121	413	101	727
Mid-west	74	71	538	138	821
West	67	67	257	77	468
North-west	125	82	193	56	456
East	67	135	1,156	279	1,637
Total	695	750	3,304	818	5,567

Source: As in Table 9.1.

since 1967, in constant 1982 prices. The average grant per project for the three groups that dominated in terms of numbers of projects — metals and engineering, wood and furniture and miscellaneous — were near enough to the overall average, so that their share of total grants was broadly similar to their share in number of projects. Two other groups — food and chemicals — received much higher than average grants per project, while another two — textiles, and clothing and footwear — received much lower grants per project. As we shall see

Table 9.3: *Industrial and regional breakdown of grant approvals, 1967–82, at constant 1982 prices*

	Total SIP grants	*Average grant per project*
Industry Group	*£m*	*£'000*
Food	17.7	55.5
Drink and tobacco	0.8	53.3
Textiles	6.0	23.3
Clothing & footwear	11.2	26.4
Wood and furniture	34.5	37.8
Paper and printing	9.6	46.4
Chemicals	11.2	78.9
Minerals	15.1	43.5
Metals & engineering	103.2	49.8
Miscellaneous	34.3	39.3
Total	243.6	43.7
Region		
North-east	22.8	50.1
South-east	22.2	39.2
Midlands	19.1	43.8
South-west	35.1	48.3
Mid-west	34.4	41.9
West	20.3	43.4
North-west	23.9	52.4
East	65.7	40.1
Total	243.5	43.7

Source: As in Table 9.1.

later, these differences are primarily accounted for by differences in capital intensity among the groups. Regional differences in average grant per project were much less pronounced, and arise mainly from the policy of giving lower capital grants as a proportion of investment in the more developed regions.

Anticipated Grant Cost per Job and Capital Intensity

From an administrative viewpoint, a key magnitude at the approval stage is the proportion of fixed asset investment to be financed by capital grants, referred to here as the "approved grant rate" (G/K). The highest permissible value which G/K can take for SIP firms is 60 per cent in designated areas. The size of approved grants is determined also, however, by the anticipated employment (L). Normally each grant approval is associated with a given number of job approvals, representing the net additional increase in employment expected once the project reaches full production. The ratio G/L indicates the anticipated capital grant cost per job, while the ratio K/L shows the anticipated capital intensity of the project. These three ratios are related as follows

$$\frac{G}{L} \equiv \frac{G}{K} \cdot \frac{K}{L}$$

The values of these three ratios are given in Table 9.4 for the SIP for each year covering the period 1967–82. We shall examine later the degree to which the anticipations are translated into actuality.

Table 9.4: *Anticipated grant ratio and capital intensity in SIP, 1967–1982, at constant 1982 prices*

	Capital grants/ fixed assets (G/K)	*Fixed assets/ anticipated jobs (K/L)*	*Capital grants/ anticipated jobs (G/L)*
	%	£	£
1967	53.6	4,393	2,357
1968	59.0	8,418	4,969
1969	56.4	9,536	5,375
1970	45.6	8,957	4,089
1971	45.7	7,459	3,411
1972	43.5	7,980	3,467
1973	39.6	8,673	3,439
1974	41.0	7,495	3,071
1975	40.6	6,967	2,827
1976	38.9	9,895	3,847
1977	40.2	7,319	2,944
1978	45.7	7,985	3,648
1979	44.5	7,818	3,475
1980	43.6	7,960	3,470
1981	39.0	7,251	2,830
1982	36.1	7,491	2,702
Total	42.3	7,897	3,340

Source: As in Table 9.1.

As may be seen from Table 9.4, the anticipated capital grant rate was near the upper limit of 60 per cent in the early years of the SIP. It then declined as the programme was extended to other areas to which a lower grant limit applied. The data on anticipated fixed assets per worker do not reveal any upward trend over the period, however, and neither is there any upward trend in the average capital grant cost per worker. This is surprising, given the presumption that over such an extended period capital intensity would be rising. Examination of the NIP indicates that a similar situation holds there.

Table 9.5 shows the same data broken down by industry group. The four industries which received the highest capital grant per anticipated job — food, drink and tobacco, chemicals and minerals — all had a high capital intensity as indicated by the amount of fixed assets per job. Equally the two industries with the lowest capital grant cost per job — textiles, and clothing and footwear — both had low capital intensity. The correlation for the ten groups between the two variables was 0.82, which is significant statistically. Although the ratio of grants to fixed assets did vary somewhat among the groups, there was only a slight tendency for the grant rate to vary inversely with the degree of capital

Table 9.5: *Anticipated grant rate and capital intensity by industry group 1967–1982, at constant 1982 prices*

	Capital grants/ fixed assets (G/K)	*Fixed assets/ Anticipated jobs (K/L)*	*Capital grants/ Anticipated jobs (G/L)*
	%	£	£
Food	43.3	9,945	4,302
Drink & tobacco	42.3	12,396	5,239
Textiles	43.3	6,331	2,744
Clothing & footwear	42.8	4,321	1,850
Wood & furniture	46.5	7,786	3,617
Paper & printing	38.5	10,012	3,859
Chemicals	32.5	13,289	4,315
Minerals	59.0	8,331	4,918
Metals & engineering	41.5	7,830	3,253
Miscellaneous	39.3	7,745	3,044
Total	42.3	7,897	3,340

Source: As in Table 9.1.

intensity, the correlation between the two variables being -0.37, not significant at the five per cent level. The regression of the capital grant cost per job on capital intensity, both measured in log values, yielded an elasticity not significantly different from unity:

$$\log G/L = 0.571 + \underset{(5.48)}{0.842} \log K/L \qquad r^2 = 0.790$$

(t value in parentheses)

This evidence suggests that capital intensive projects were more costly in terms of anticipated grant cost per job, and that this pattern was only slightly modified by variation in the grant rate. These findings indicate that the scale of grants was determined much more by the amount of investment than by the prospective number of jobs, with a much stronger tendency towards a common grant rate as a proportion of fixed assets rather than towards a common grant amount per job. There was, however, a slight tendancy towards a relatively higher grant rate where the number of jobs was relatively high in relation to investment. These findings are broadly in line with those of McAleese (1977) for NIP industry over the period 1952-74, where there was a strong positive relationship between capital intensity and capital grant cost per job, and only a weak negative relationship between the capital grant rate and capital intensity.[38]

Finally, we compare in Table 9.6 the values of the variables in recent years in the SIP with those of the NIP and the Enterprise Development Programme (EDP). The grants data here include those for training, rental of factory space, etc., and differ from those given above, which related to capital grants for fixed assets only. Despite the fact that the grant rate is considerably higher for SIP than for NIP projects, the anticipated grant cost per job is nearly twice as great in the latter. The reason is that the capital intensity of NIP projects is about three times that of SIP projects. EDP projects have much the same degree of capital intensity as SIP projects, but because the grant rates are on average more generous in the former, the anticipated grant cost per job is greater for EDP projects, though not as high as for NIP projects. It emerges therefore that, at least as far as anticipations go, the grant cost to the state of generating industrial employment is lowest for SIP projects. We now turn to examine the relationship between anticipations and actuality.

[38]McAleese's figures related to actual amounts of grant payments, investment and jobs. It might also be noted that two econometric studies by Ruane (1976, 1978) were inconclusive as to the relationship between capital intensity and the grant rate — in that the findings of the two studies conflicted.

Table 9.6: *Anticipated total grant rates and capital intensity in different IDA programmes, 1980–82*[a]

	1980	*1981*	*1982*
Grants/fixed assets		%	
SIP	50.9	51.3	52.5
NIP	30.0	29.6	33.3
EDP	60.2	58.2	76.2
Fixed assets/anticipated jobs		*(Constant 1982 £)*	
SIP	8,655	8,114	8,192
NIP	24,443	28,336	24,854
EDP	8,144	6,900	8,816
Grants/anticipated jobs		*(Constant 1982 £)*	
SIP	4,406	4,159	4,301
NIP	7,342	8,398	8,284
EDP	4,899	4,016	6,721

[a]The grants data here include other grants besides those for acquisition of fixed assets.
Source: IDA *Annual Reports*. Data for small indigenous industry in the mid-west region catered for by SFADCo are not included here.

2 *Conversion Rates*

While the approvals data are of interest in indicating the nature of commitments, greater interest centres on the degree to which these commitments are realised. Assessment of the degree to which anticipations are converted into actuality, however, is not at all a straightforward matter. In the case of each grant approval, a period of time elapses between the date of approval for grant aid by the IDA and the date of first grant payment. The period of time involved will vary from case to case. In many cases, the grant is paid in instalments over a number of months or even years. In some cases, the amount which was approved may never be paid in full, if at all. The investment plans drawn up and agreed with the IDA on the occasion of an application for an approval may be radically changed subsequently, due to changed business conditions or other factors. Stages of a grant payment may be cancelled, deferred or even brought forward. In many instances, the payment of later stages of a grant may depend on the attainment of satisfactory performance by the establishment in the initial period of expansion following an approval.

In the case of jobs, the position is even more complex. Even if the full amount of the grant is paid out quickly following approval, this only means that the fixed assets have been installed, and it does not necessarily mean that the anticipated employment has materialised. In fact, even in the most favourable cases, employment generally builds up more slowly as production expands towards capacity levels, and even then the employment at capacity level may be less than expected at the approval stage. In less favourable cases capacity output may never be reached, or having been reached may decline again so that some of the employment gain is only temporary. There is also the difficulty, referred to in Chapter 8, of estimating how much actual employment should be associated with the grant payments. This issue is complicated further by reason of the fact that before the build up of employment from an initial grant is completed, a new grant-aided expansion may be taking place.

Conversion Rates for Grants and Jobs

It will be clear, therefore, that any measure of the rate of conversion of approvals into actuality is bound to be somewhat arbitrary. What we have done here is to examine the cohort of firms first approved for a SIP grant in the years 1973–77 which had received at least one capital grant payment in any of the years 1973–78. There were 377 such establishments, of which 228 were already in production at the start of the year in which they first received SIP approval (referred to here as "existing" establishments) and 149 were not ("new" establishments). The conversion rate for capital grants is measured as the ratio of cumulative capital grant payments over the years 1973–78 to cumulative capital grants approved over 1973–77, all measured at constant 1982 prices.

In the case of jobs, the actual employment attributed to the SIP is measured as the net difference between employment in each establishment on 1 January of the year of first approval for a SIP grant and its employment on 1 January 1979. Job losses in establishments that had closed or contracted by the terminal date are counted with a negative value. No account is taken of temporary employment gains that were not in being on 1 January 1979, or of employment gains realised after that date. The conversion rate for jobs is the ratio of the sum of the specified employment changes to the total of job approvals in these establishments in the period 1973–77.

Table 9.7 gives the resulting data. Of the total of £22.2m. (at 1982 prices) approved for the establishments from 1973–77, £12.5 million had been paid out by the beginning of 1979, involving a conversion rate of 56 per cent. The rate was slightly higher for existing projects than for new ones. The jobs conversion rate was considerably lower at 44 per cent. Moreover, there was a much wider gap in the rates for existing and new establishments, with the jobs rate being considerably better for the new establishments. For new establishments for the jobs conversion rate was much the same as the grants conversion rate, whereas for existing establishments, the former rate was only half the latter. The lower grants conversion rate for new establishments is explicable on the grounds that establishments

Table 9.7: *Conversion rates for capital grants and jobs in establishments first approved in 1973–1977*

		Capital grants (1982 prices)			*Employment gains*		
	No. of Establishments	*Anticipated*	*Actual*	*Conversion rate*	*Anticipated*	*Actual*	*Conversion rate*
		£m.	£m.	%			%
Existing establishments	228	9.7	5.9	61.0	3,288	1,088	33.1
New establishments	149	12.5	6.6	52.2	3,634	1,961	54.0
Total	377	22.2	12.5	56.0	6,922	3,049	44.0

Source: IDA *Project Information File.* For explanation of the concepts, see text.

already in existence are better placed to complete their investment quickly, whereas new establishments may encounter more problems in starting. One might expect the same factor to hold in the case of jobs. Once an investment in a new enterprise does get off the ground, however, then a certain minimal amount of new employment must take place if there is to be any production at all. In established firms, however, this is not necessarily so, since some of the existing labour may be re-allocated to operate the new plant. This may explain, in part at least, why the jobs conversion ratio is so much lower in existing firms. The point will be considered further, however, in the concluding section of this chapter.

In calculating the conversion rates in Table 9.7, we included firms which had closed down by 1979, but excluded projects which had not got any payment at all by this date or which were cancelled. If we were to include cancellations and deferments, then the capital grants conversion rate would fall from 56 to 51.5 per cent, while the jobs conversion rate would fall from 44 to 39.1 per cent. If, on the other hand, we excluded not only cancellations and deferments but also establishments that had gone out of business then the conversion rate for the surviving establishments would be 56.5 per cent for grants and 56.1 per cent for jobs. Since these two figures are virtually identical, this might suggest that closures accounted for the disparity noted earlier between the grants conversion rate and the jobs conversion rate. That conclusion, however, would be misleading as regards existing establishments. When closures are excluded the new establishments had a substantially better jobs conversion rate (67.7 per cent) than their grants conversion rate (52 per cent), whereas there was still a wide gap for the existing firms — 44.9 and 61.8 per cent, respectively.

In calculating the conversion rates above, we used a fixed terminal point (1979) to measure the rate of conversion for projects approved at different points of time in the period 1973–77. It was also possible, however, to look at the jobs conversion rate within two years of approval for yearly cohorts of approval. The conversion rates were much higher for the 1976 and 1977 cohorts than for those of the years 1973–75, reflecting the effects of the recession on the latter. For the 1973–75 projects, less than one-quarter of anticipated jobs were realised two years later, whereas for the 1976–77 projects the figure was 45 per cent. For all years from 1973 to 1977 taken together, the jobs conversion rate within two years of approval was 34.8 per cent.

Examination of the employment experience in establishments which were already in existence before receiving SIP assistance showed that in general their employment was increasing before as well as after being approved. This raises the question, already mentioned earlier, as to whether they would have gone on growing anyway without SIP assistance. To throw some light on this, we examined the growth rate before and after receiving SIP approval. The weighted average annual growth rate prior to year of first approval worked out at 4.5 per cent per annum, and 10.5 per cent per annum subsequently. While this does not prove that SIP assistance was responsible for the faster growth rate, especially as the later years covered were ones of more buoyant activity overall, nevertheless it is reassuring that there was a strong acceleration in growth.

Undoubtedly, however, many IDA assisted small projects which existed at the year of first approval were selected for grant aid partly on the basis of a good employment record prior to approval. Thus it is not surprising that some positive employment growth was occurring in these projects before approval. Yet, the corollary of this is that employment might have continued to grow irrespective of any grant assistance. True, some establishments would have gone out of business, contracted or expanded more slowly in the absence of aid, but it would be stretching things to say that no net employment increases would have materialised without SIP assistance. To the extent to which some employment growth would have occurred irrespective of grant aid, estimates of grant cost per job created are understated, and this should be borne in mind in interpreting the figures which follow — not only for SIP projects, of course, but for other grant programmes also.

Actual Grants Rate and Cost per Job

For the same cohort of firms — those approved for the first time in the period 1973-77 and receiving at least one capital grant payment before the end of 1978 — Table 9.8 presents data for the anticipated and actual values of the capital grants rate (G/K), capital intensity (K/L), and capital grant cost per job (G/L). The actual capital grant cost per job for the full cohort was 27 per cent greater than anticipated. Since the actual and anticipated grant rates are very similar to each other, the disparity in grant cost per job is chiefly due to the higher than

Table 9.8: *Comparison of anticipated and actual grant rate and capital intensity in SIP cohorts*

	Capital grants/ fixed assets (G/K)		*Fixed assets/jobs (K/L)*		*Capital grant cost per job (G/L)*	
	Anticipated	*Actual*	*Anticipated*	*Actual*	*Anticipated*	*Actual*
	%	%	*Constant 1982 £*		*Constant 1982 £*	
Existing establishments	39.0	37.3	7,505	14,288	2,923	5,390
New establishments	40.9	39.1	8,390	8,500	3,434	3,321
Total	40.0	38.4	7,969	10,566	3,191	4,061

Source: Derived from IDA *Project Information File*. The cohort of firms is the same as in Table 9.7.

anticipated level of capital intensity. No such disparity arose in the case of new establishments where the actual values of the three variables were very close to those anticipated. For existing establishments, however, actual capital intensity of projects was nearly twice as great on average as that anticipated, chiefly due to the low jobs conversion rate mentioned earlier. The results show that while the anticipated capital grant cost per job in new projects was nearly 20 per cent higher than in existing projects, in actuality it turned out to be about 40 per cent less.

Comparison with Other IDA Programmes

McAleese (1977) reported a jobs conversion rate of 70 per cent by end 1974 in NIP industries which were approved in the period up to 1973. At first sight this figure seems very much higher than the jobs conversion rate of 44 per cent given above for the SIP cohort of establishments. There are a number of factors however that vitiate the comparison. McAleese excluded not only cancellations but also closures. Moreover, he excluded existing establishments in which the performance would not have been as good as in new firms. For the SIP cohort, if one confined it to new firms and excluded those firms that closed, then the jobs conversion rate would have been 68 per cent.

There is a further difficulty, however, in comparing our results with the figures of McAleese (1977) in that the two periods were rather different, and in particular that employment experience generally was less favourable in the later period to which our data relate. A contemporaneous comparison is afforded by the data in NESC (1982) which are given in Table 9.9, adapted to constant 1982 prices for comparison with the SIP data. It is clear that the conversion rates are higher and the actual grant cost per job lower for SIP firms than for other indigenous

Table 9.9: *Comparison of conversion rates and grant cost per job in SIP and other IDA programmes*

	Grants conversion rate	*Jobs conversion rate*	*Capital grant cost per job*	
			Anticipated	*Actual*
	%	%	*Constant 1982 £*	
(1) SIP cohort (1973–1977)	56.0	44.0	3,191	4,061
(2) IDA indigenous industry (1973–79)	46.1	16.7	5,694	15,689
(3) IDA foreign industry (1973–1980)	21.0	23.7	10,375	9,201

Sources: (1) Derived from IDA *Project Information File,* (2) and (3) adapted from NESC (1982). The data in (2) and (3) include grants under all the main development programmes (NIP, SIP, R & D, etc.) but not re-equipment grants. Most SIP firms would be included as part of (2).

firms or for foreign firms.[39] While the performance in indigenous industry as a whole is poor, it would obviously be much worse if the SIP element were excluded.

It might be argued that the impact of the recession in the late 1970s has borne particularly heavily on employment in larger firms, and that in more normal conditions the grant cost per job in other industry would not be so high relative to SIP projects. The figures in McAleese (1977) for the earlier and more buoyant period do indeed show a much lower actual capital grant cost per job for NIP projects than those reported in Table 9.9. However, McAleese's figure, converted to 1982 prices, at £7,321 is still much higher than the figure for SIP projects in Table 9.9. It seems clear then that in so far as grant cost is concerned, the SIP has been a relatively cheap method of creating jobs.

3 *Closures*

The data in Chapter 8 relating to employment losses due to establishment closures suggested that SIP firms did not record a particularly high failure rate compared to firms in other IDA programmes or in manufacturing generally. In this section the pattern of closures in SIP establishments is examined in greater detail in relation to projects approved up to 1978.

Of the 1608 projects approved for a SIP grant between 1967 and 1978, 85 were not proceeded with and a further 50 began only by 1 January 1980 or later. Of the 1,473 projects approved in the years 1967–78 which had been in production at least one year prior to January 1980, 265 had closed by 1 January 1980,[40] or 18 per cent. In a previous study of NIP projects which received grant payments between 1960 and 1973, O'Farrell (1975) found that 16 per cent had closed by June

[39]It is not clear if the NESC figures refer only to capital grants, as is the case with our SIP figures, or whether they also include training, leasing and other grants. If these were included for the SIP, they would push up the grant cost per job by about one-quarter, but would still leave it well below the figures for indigenous and foreign industry.

[40]Since we did not have precise closing dates, we had to identify them by the employment records available only at 1 January each year since 1973. An establishment is regarded as having closed in the year preceding that in which it ceases to have a positive employment record. In a very small number of cases an establishment may have temporarily ceased employment for a year or more but recommended by January 1980: these cases are *not* included as closures. Establishments that ceased employment in 1979 or in earlier years but only recommenced after 1 January 1980 *are* treated as business closures, since we do not have available a record of their re-establishment. There are also a small number of cases where an establishment changed location or industry branch which show up here as closures because of the way establishments are recorded. For years prior to 1973, when no Employment Survey was carried out, there is a record available of the number of closures, but not of the dates at which establishments closed or the employment levels at time of closure.

1973. For a slightly later period covering all firms that had received NIP grants up to end 1974, McAleese (1977) found that 20 per cent had closed by end 1975.

None of these figures, however, can be interpreted as a closure *rate* that can be compared with each other. The reason is that older firms will tend to have a high cumulative proportion of failures simply because they are longer in existence, so that the crude proportions shown are influenced by the length of the time period covered and the distribution of the projects over time. Sometime closer to a standardised annual average closure rate may be got by considering the ratio of closures in each year to the stock of plants in that year and then taking the mean of the annual closure rates.

Table 9.10 shows the annual number of closures each year for 1973-79 for establishments approved in the period 1967-78.[41] The mean of the annual closure rates in the period 1973-79 works out at 3.5 per cent. This is higher than the mean annual closure rate estimated by O'Farrell (1976) for NIP projects of 2.6 for the years 1966-72 and 2.9 for the years 1967-72. These periods were more buoyant economically, however, than the years following 1973 and O'Farrell

Table 9.10: *Annual closure rates 1973-79 for SIP establishments approved in the period 1967-78*

	Number of establishments			*Employment*		
Year	*(1) No. of SIP estbls. in employment*[a]	*(2) No. of closures*	*(3) (2) as % of (1)*	*(4) Employment in all estabs. in (1)*	*(5) Employment in closures*	*(6) (5) as % of (4)*
1973	624	13	2.1	8,033	66	0.8
1974	711	27	3.8	9,756	258	2.6
1975	775	35	4.5	10,636	304	2.9
1976	832	32	3.8	11,721	372	3.2
1977	945	28	3.0	13,939	345	2.5
1978	1,131	51	4.5	17,297	700	4.0
1979[b]	1,163	32	2.8	19,117	351	1.8

[a]The establishments are those which received SIP approval in the period 1967-78 and which had been in employment at least one year prior to 1 January 1980. Data for establishments already in existence prior to receiving SIP approval are included only from the year in which they first received approval.

[b]There were 14 other closures in 1979 in respect of firms first approved in 1979 which are not included here.

Source: IDA *Annual Employment Survey 1980*.

[41]It is not possible to show closures in earlier years than 1973, since employment data are not available and only the number of closures, and not their date of closure, is available for 1967-72.

and Crouchley (1983) reported a considerably higher rate for the period 1973–81. They also found that the proportion of the 1973 stock of plants which closed by 1981 was somewhat higher for NIP than for SIP plants.

Table 9.10 also gives associated employment data. The average of the annual closures rates for the period 1973–79 in terms of employment works out at 2.5 per cent,[42] considerably lower than the rate measured in terms of number of establishments. A corollary of this is that the mean size of closures in the period 1973–79 in terms of employment was lower than in all SIP establishments — 11 as against 14.5. These findings are in line with those of O'Farrell (1975) and McAleese (1977). The employment loss, of course, is measured by reference to the level in the year of closure, and reflects the likelihood that closure was preceded by a decline in numbers engaged. McAleese (1977) noted that while this type of statistical distortion goes part of the way to explaining the link between small size and closure rates, it did not constitute a full explanation. Some support for this view is given by the SIP closures in that the average employment in 1973 of the establishments then in existence which closed in the period 1973–79 was already below the overall average in 1973, 11.3 as against 13.3. For all closures, however, there was an upward trend over time in their mean size. While this partly reflects the rising mean size of establishment for all SIP establishments, the relative difference between the mean size of closures (in the year of closure) and all SIP industry in the same year narrowed over the period 1973–79.

Industrial Classification of Closures

There was considerable variation in the proportion of establishments which closed in different industrial groups. In terms of number of establishments, the highest proportions were experienced in clothing and footwear (26.1 per cent), textiles (23.9 per cent) and food (22.8 per cent) — compared to the overall figure of 18 per cent. The lowest proportions were in metals and engineering (15 per cent) and minerals (16.9 per cent).[43] As regards employment losses, the mean establishment size in the year of closure in all groups which had more than five closures in the period 1973–79 was not far off the overall average of 11 with the exceptions of clothing and footwear (19.8) and minerals (2.4). Thus clothing and footwear had not only a very high proportion of closures in terms of number of establishments, but the establishments involved were relatively large. Of the total employment losses due to closures in the period 1973–79 in all SIP estab-

[42]This is the same as the figure estimated for indigenous SIP firms in NESC (1983), Table 3.5, for the period 1973–80. For all SIP firms, including foreign SIP firms, however, the NESC study estimated an average annual rate of 3.0.

[43]There was also a very low proportion of closures in chemicals (7.3 per cent) and paper and printing (12.8 per cent) but the absolute numbers of all establishments involved in these groups were small so that not too much can be read into the figures.

lishments approved in the period 1967–78, this group accounted for 18 per cent, more than double its share in SIP total employment. At the other extreme, minerals accounted for only a little over one per cent of job losses due to closures, well below its share in total SIP employment. Metals and engineering had the highest share of total job losses due to closures, 34 per cent, but this figure was still well below its share in total SIP employment.

Table 9.11 summarises the different industrial experience in regard to closures in SIP establishments, by relating the number of job losses due to closures in the period 1973–79 to the average of the total levels of SIP employment on 1 January 1973 and 1980. As before, the closures are in those establishments approved in the period 1967–78. Clothing and footwear stands out as the group with by far the highest proportion of job losses from closures, due no doubt to the heavy foreign competition suffered by most segments of this industry in this period, as well as other factors like fashion changes combined with ease of entry. Minerals and chemicals had an exceptionally low proportion of losses due to closures. The low figure for drink and tobacco is not of great significance because of the very small numbers involved. As suggested above, the largest SIP group, metals and engineering, also emerges with a low proportion of jobs lost through closures.

Grants to Closures

The total grants approved in the period 1967–78 for establishments which had closed down by 1 January 1980 amounted to £10.9m. in constant 1982 prices.

Table 9.11: *Employment losses due to SIP closures 1973–79 as a percentage of total SIP employment by industry group*[a]

	%
Food	14.0
Drink and tobacco	3.3
Textiles	16.9
Clothing & footwear	30.8
Wood and furniture	12.0
Paper and printing	10.8
Chemicals	9.4
Minerals	2.6
Metals & engineering	10.9
Miscellaneous	16.5
Total	13.2

[a]The precise figures used are explained in the text.
Source: IDA *Annual Employment Survey, 1980.*

This is equivalent to 11.6 per cent of the total approved for all SIP establishments in the period. Since the share in grant approvals is less than the proportion of establishments closing down, it is clear that the average size of project among closures, as measured by average size of grant, was less than for all SIP projects. The anticipated capital intensity and grant cost per job were both also slightly lower in closures than in all SIP projects.

Total grants actually paid to these closures amounted to £6.7m. in constant 1982 prices. In order to compare the conversion rate for closures with a similar group of surviving establishments, we used the 1973–77 cohort referred to earlier. For all the establishments involved the capital grants conversion rate was 56 per cent. There was very little difference between the rates for closures and surviving establishments — 54.5 per cent and 56.5 per cent, respectively.

O'Farrell and Crouchley (1983) found that for Irish industry in general grant-aided firms had a lower probability of closure than those not grant-aided. While recognising that this might be due to the ability of the development agencies to pick winners, the authors argued however that the more important explanation lay in the effect of the grant in enhancing viability in the early vulnerable years when cash flow is often a serious problem.

4 *Conclusions*

As with other IDA programmes, the SIP shows a significant divergence between anticipated and actual performance. Over half the jobs approved in 1973–77 had not materialised by 1979. The rate of conversion of grants was somewhat greater, however, so that actual grant cost per job worked out higher than anticipated. These results indicate a need for discount the approvals data by a significant factor if they are used as a guide to future actual performance. In fact the IDA no longer use job approvals as a measure of activity in small industries, the target in this regard being first-time jobs actually created.

Nevertheless, on all these counts, the SIP experience was more favourable than for NIP firms in general and NIP indigenous firms in particular. The findings suggest that a great proportion of SIP approved jobs were converted into actuality and at a much lower grant cost per job, while the rate of closure was not greatly different. True, SIP projects take up more IDA staff time than larger projects, but while we did not have estimates of this, it would scarcely offset the wide margin in favour of SIP projects in terms of grant cost per job. It may be that the superior performance in small industry is not due to smallness, but rather to the more directive approach adopted by the development authorities in relation to small firms. If so, this would point to the adoption of a more directive approach for large firms. This argument will be considered further in the next chapter.

In evaluating such findings, however, it should again be borne in mind that there is a measure of interdependence between the different components of industry. Without the development of larger firms, many of the new small firms would not be possible. Some of them directly supply inputs to the larger firms, while many others depend on the markets arising from the increased activity and incomes generated by larger industry. Only a minority of the small firms would be able to enter and survive on the basis of export markets. In that sense small industry is the more dependent component. But though large firms might be able to get their sub-supplies from abroad, nevertheless, the existence of efficient sub-supply activities near to hand can also be a considerable help in improving the environment for developing larger industry. Hence in considering overall industrial strategy, the interrelationships between different programmes must be considered and the relative cost of jobs in different programmes cannot alone be the decisive factor.

Among SIP assisted establishments, new entrants performed better than established firms in relation to converting anticipated jobs into actuality, and their realised capital grant cost per job were lower. It does not neccessarily follow, however, that greater concentration on securing new entrants would be the better avenue to industrialisation and job creation. In existing establishments, the new investment in part serves to increase the viability of existing jobs — a function which by definition it cannot perform for new establishments. In effect the new investment in existing firms may effectively be used in part to equip the existing workforce better. While this would imply a tendency to greater overestimation of anticipated new jobs in existing establishments than in new ones, it would also imply that the employment contribution made by the investment, taking account of job preservation, would be greater than indicated by the net change in employment.

Moreover, the objective of the SIP is not simply to create many viable small firms *per se*, but also to develop some of them into larger enterprises. In facing this challenging task, the difficulties involved may not be substantially different in new firms once they have reached a certain size from those already encountered in developing longer-established firms. True, relatively more of the older firms may have become entrenched in their ways, or be located in areas or activities with low prospects for growth. Nevertheless, for new as well as for older firms, only a minority can be expected to have major growth potential, and the key to success lies in identifying and building on these, whether they be old or new. It is difficult to see how this could be done other than in the light of experience. One of the important side benefits of the SIP is that it can provide the development authorities with a great deal of specific practical experience of the establishments most likely to be suitable for larger-scale development.

Part III

POLICY CONCLUSIONS

Chapter 10

POLICY IMPLICATIONS FOR IRELAND

Policy for small industry should properly be considered in the context of overall industrial policy, which in turn must be framed in the light of general economic conditions. We begin this chapter, therefore, by outlining briefly the economic environment in which Irish industry is likely to have to operate, and the likely shape of overall industrial strategy. In Section 2 we consider the strategy appropriate to the development of small industry. The policy instruments that might be used to give effect to this strategy are considered in Section 3, and some concluding comments are made in the final section.

The environment for industrial development is strongly influenced by general policy measures as well as those specific to industry itself. These general measures relate to such matters as the improvement of development facilities (e.g., education, infrastructure), macro-economic policies affecting profitability (e.g., incomes policy), the influence of taxation on risk-taking and work effort, the rewards for saving, the ease of access generally to investment funds, and even the climate of social attitudes towards risk and failure. There is always a temptation to pick on industry-specific measures as a palliative for failure to tackle broader environmental deficiencies, which may arise partly from defective general policies.

Clearly the foregoing issues are important for successful industrial development. But there are some who would go so far as to state that they are sufficient, and that once the overall competitive environment is "right", there is no need for further industry-specific policy measures. Historical experience indicates, however, that few of the advanced countries developed their industrial base without selective measures in favour of manufacturing, at least in the earlier stages of their development; and the same holds true for the successful newly industrialising countries (NICs).[44] Ireland is still at a relatively early stage of

[44]In regard to the East Asian NICs, see for example Fransman (1984) and other articles in the same Bulletin.

industrialisation, and as a small peripherally-located, country seeking to do so in free trade conditions, it faces considerable "entry barriers" which are unlikely to be overcome without specific developmental measures. The really crucial issue seems to us not whether the State should intervene selectively to encourage manufacturing development, but rather the quality of the interventions and the effectiveness with which they are implemented.

1 *Overall Industrial Strategy*

In the years ahead the growth of domestic and foreign markets is likely to be sluggish. A satisfactory growth of manufacturing output cannot be achieved, therefore, simply by relying on maintaining the same share of growing markets: it will be necessary to aim at increasing market share, particularly in export markets. This in turn implies that progress depends on exploiting more effectively whatever competitive advantages Ireland had, and in improving these competitive advantages. Moreover, the aggregate volume of mobile international projects may be reduced, and there is likely to be intensified international competition for such projects. These circumstances suggest the need to concentrate relatively more than in the past on the development of indigenous enterprise — as, indeed, is now the stated aim of industrial policy. The two factors that contributed so much to the industrialisation of Ireland in the 'sixties and early 'seventies — buoyant demand and a large influx of foreign enterprise — are likely to be much weaker in the decade or so ahead, so that Ireland must look more to greater export market penetration by indigenous companies.

Yet, though the prospects for growth in manufacturing output and employment are less favourable, the need for such growth is in no way diminished. Recent population and labour force projections put the growth of the labour force for 1986-91 in the range 11,000 to 17,000 per annum (CSO, 1985). The former figure is based on a "high" emigration figure of 15,000 per annum and the latter on a "low" figure of 7,500 per annum. If unemployment, therefore, were not to rise above the present intolerably high levels (224,000 in May 1985), then an increase in employment of 11-17,000 a year would be needed on average over the next five years: indeed, without emigration the required increase would probably be in excess of 20,000 a year. Moreover, since agricultural employment will almost certainly fall by about 3,000 per annum, the required increase in non-agricultural employment would be even greater.

It has been argued, however, that the relationship between output growth and employment growth in manufacturing is likely to be less favourable to employment in future, and that industrial strategy would be more successful if it placed primary emphasis on wealth creation rather than on employment creation in manufacturing (White, 1983). The basis of this view is that what matters for business is long-term viability, and that this depends primarily on profitability. Manufacturing profitability, of course, is not necessarily inconsistent with employment creation in manufacturing, but it is argued that the emphasis on the latter deflects from the former. It is implicit in this view that a wealth creating strategy would probably be far more capital intensive, with less employment created *directly* in manufacturing. It is claimed, however, that in the long run the stretegy would be more favourable nationally to employment, in that it would increase the demand for labour in private services and provide more tax resources to finance employment in the public sector.

There is little doubt but that the employment growth/output growth relationship in Irish manufacturing has diminished since 1979. In the period 1973–79 when manufacturing output grew by 5.1 per cent per annum, employment rose by 0.7 per cent per annum; but from 1979–84 when output grew by 4.3 per cent per annum, employment fell by three per cent per annum. In the year 1984 itself, even though output rose by over 13 per cent, employment still fell. To some extent this latter figure may be due to the lagged effect of employment in response to recovery in output after a depression. But the major reason for the poor employment response to output growth in recent years is that most of the growth in output is coming from the new technology industries which have exceptionally high levels and growth of productivity — at least as conventionally measured. The office machinery and data processing industry, for example, had a level of net output per head in 1981 of about two and a half times the rest of manufacturing, and its measured productivity grew by 26 per cent per annum from 1979 to 1984. If output growth were to continue to be heavily concentrated in new industries with a low direct labour content and little backward linkages, then high output growth might yield only a small direct employment gain in manufacturing. Furthermore, if as seems to have been the case, a high proportion of the non-labour value added were to flow out of the country, then the impact on the economy of the output growth would be further diluted.

It has to be admitted that, no matter which strategic approach is adopted, manufacturing itself can only provide directly a minority of the jobs needed to keep unemployment from rising further. A major study of employment policy in Ireland (Conniffe and Kennedy (eds.) 1984) accepts that under any realistic scenario, the majority of the jobs will have to be found in private services and in publicly-funded employment, whether in the regular public sector or in special employment schemes, or in the "third sector". The third sector is basically a partnership of public, community and voluntary organisations to meet certain community needs, generally in the social fields, such as day care, housing, local health and rehabilitation centres, etc. Finance is provided by the State, either on a once-off basis or by way of continuing subsidy, but the management of the services and recruitment of staff are left to the groups themselves. Not all third-sector activities, however, need be confined to social services. Local community enterprises may succeed in some economic activities — even in areas of manufacturing — in which conventinal private enterprise has shown no interest, and may be financially viable on the basis of incentives no more costly than those given to private enterprises.

Manufacturing can make a major indirect contribution to employment creation in other sectors by relaxing the balance of payments and public finance constraints. Nevertheless, Conniffe and Kennedy also take the view that some direct employment increase must be sought in manufacturing itself. The share of manufacturing employment in total non-agricultural employment is much lower in Ireland than in most OECD countries; and lower still than the shares prevailing when those countries were at a comparable stage to Ireland in the industrialising process. It is highly unlikely that Ireland will progress satisfactorily unless the present share were to be at least maintained. If total non-agricultural employment is to grow at 20,000 a year, then maintaining the manufacturing share would require a growth in manufacturing employment of about 5,000 a year. Indeed it could be argued that a faster rate is needed to bring Ireland's manufacturing share closer to other developed countries.

Exclusive reliance on raising the value added of manufacturing, without any direct employment gain, would place an enormous burden on the "transfer" process by which resources generated in manufacturing give rise to jobs in services. Increased manufacturing value added must accrue initially in the form of incomes — wages or profits — to the persons engaged directly in manufacturing. In a very open economy, a high proportion of these incomes is likely to transfer overseas, either in repatriated profits or in spending on imports. Higher taxes could in principle be used to fund more employment, but there is strong resistance to increasing tax rates. Moreover, the political will to allocate large public funds to the development of manufacturing derives much of its impetus from the fact that the industrial programme is seen as directly contributing to the social goal of employment. In the light of these considerations, we take the view that an overall strategy which led to a situation where little employment resulted directly in manufacturing would not be sustainable.

In practice, however, the dichotomy between the two strategies — wealth creation and employment creation — is far from absolute, and it is not a question of choosing one rather than the other. The achievement of rates of growth of output and employment in manufacturing of the order mentioned above will clearly require favourable conditions for profitability. Moreover, it is not necessary to have the same strategic approach in relation to all aspects of manufacturing. Where, for example, a project with high capital intensity manifestly yields benefits favourable to employment in other branches of manufacturing or in the non-manufacturing sectors of the economy, then it should not be ruled out simply because it adds little directly to employment. Throughout this study we have stressed the interdependence, actual and potential, between different sectors of manufacturing, and in particular between large industry and small. Given the established record of small industry as a significant contributor to employment creation, the foregoing considerations would point to the development of small industry as an interdependent feature in overall industrial strategy that would have particular relevance to job creation.

The Telesis Strategy

In the late 1970s, the National Economic and Social Council, with the Government's endorsement, sponsored a series of studies of industrial development policy. In the context of future policy directions, the key study is the Telesis Report (1982), which proposed certain modifications to the prevailing industrial strategy and the addition of new strategic dimensions. The general intent of the Telesis proposals, which were largely endorsed by the Council, was to concentrate on the development of indigenous industry and those foreign companies with characteristics relevant to the long-term strength of Irish industry. The chief emphasis would be on producing traded goods in enterprises that would be large enough to serve world markets. Non-traded activities would be less favoured than heretofore, except in the case of high-skilled sub-supply activities. As regards type of industry, the prime focus would be on "complex factor cost businesses" where the key to competitive success lies in skill levels, innovation, marketing, etc., rather than in low wages or in logistic factors.[45]

In developing a suitable corporate shell to undertake a selected activity, Telesis favoured a more directive approach on the part of the development agencies. Existing companies would be encouraged to rationalise and combine, joint ventures might be arranged, or holding companies formed. There would be a more active dialogue between government policy-makers and large companies about investment plans, and sticks as well as carrots might be used. Policy instruments would be designed to address specifically whatever cost penalties had to be overcome in each selected enterprise — whether they related to product or process technology, overseas marketing, skill development, working or fixed capital, etc. The instruments would also seek to develop capabilities in firms rather than in the development agencies.

The White Paper on Industrial Policy

The Government White Paper (1984) on *Industrial Policy* outlines the Government's intentions about the future direction of industrial policy. The

[45]A well-documented case for the broad approach advocated by the Telesis Report is made by O'Brien (1985) in relation to electronics and other high-technology activities.

White Paper established an objective doubling output in the next decade, and a growth in manufacturing employment of between 3,000 and 6,000 a year. Many of the measures recommended to achieve this goal are in line with those recommended in the Telesis Report. Thus, for example, the industrial incentives are to be directed more selectively towards firms producing traded goods, there is to be a gradual shift in resources from fixed asset grants to technology acquisition and export marketing, greater linkages are to be encouraged, and a National Development Corporation is to be established which would, *inter alia*, initiate new projects on its own or invest on a selective basis in the development of structurally-strong Irish firms.

There are, however, important differences between the two documents, as least in emphasis. This appears early on in the view taken in the White Paper of the Government's role in industrial development, which is stated to be the giving of incentives and advisory services, the provision of infrastructure, the encouragement of a favourable economic and fiscal environment and the development of viable job creating projects in the commercial State sector (p. 5). This general statement seems to stop well short of the "hands on" approach by public agencies recommended by Telesis for the building of strong indigenous industry. The disparity is confirmed in several detailed aspects of the two strategies. Thus, whereas the Telesis Report recommended a goal of raising the proportion of funds allocated to indigenous export or skilled sub-supply firms from less than 40 per cent over the last 10 years to 75 per cent by 1990, no such target is included in the White Paper. Second, the White Paper took the view that in the aggregate "the state of Government finances does not allow for any real expansion of expenditure on job creation" (pp. 3–4); while at firm level, the aim is that "the overall amount of money which firms will receive in grants will be at the same average level" (p. 41). Given the Telesis view, which seems well-founded, that job creation in indigenous firms would initially be more expensive, it is difficult to see how a substantial shift towards development of indigenous industry would arise, given the constraints set by the White Paper on funding. Third, the White Paper seems to envisage a less active role than Telesis in securing the rationalisation of firms to achieve appropriate scale by, for example, making grants conditional on inter-firm co-operation or grouping.

Of course, as has been pointed out before (e.g., Kennedy, 1982), the Telesis recommendations are fairly general and much work would need to be done before the strategy could be translated into an operational blueprint. Only time can tell, therefore, to what extent the Government will in practice seek to implement it. In our view, there is much merit in the Telesis strategy for the development of large indigenous companies. Irish manufacturing has relatively few large firms and their average size is small by reference to international experience in developed countries. In seeking to enlarge the industrial base in Ireland, enough has been said in earlier chapters to indicate that small firms cannot form the only, or indeed, the main component. There *are* major economies of scale in production; the export marketing capability of individual small firms is limited; innovations, as distinct from inventions, can generally be exploited fully only by large-scale firms; while large firms are often necessary to provide markets for small — either directly through subcontracting, or indirectly by increasing incomes. Thus, it seems sensible to stress, as the strong right arm of industrial strategy, the building of a selected number of large companies that would eventually be able to engage on their own in the full range of functions needed for success in world markets. We now turn to the question whether there is scope for a second arm focusing on small firms.

2 *Strategy for Small Industry*

Discrimination in favour of small industry is warranted only if particular benefits to society can be expected. The mere fact that small-scale firms suffer various economic handicaps is not sufficient to justify policy measures to remove such handicaps. Thus, for example, the fact that small firms have to pay more for their purchases is not in itself a cause for concern where it simply reflects the fact that smaller orders are more costly to handle. Though it is a differential cost penalty to the small firm, it does not provide a justification for intervention unless corresponding differential benefits can be expected to accrue to the society from such intervention. As we have seen in the last chapter, many countries have traditionally taken the view that no special discrimination in favour of small industry is warranted, though even in these countries special measures are taken to remove "artificial" disadvantages facing small firms that are not justified by genuine diseconomies of scale, and the range of such special measures has increased considerably in the last decade.

Telesis and Small Industry

In so far as overall industrial strategy is concerned the Telesis Report clearly rejected the approach of sole reliance on general policy measures and minimal interference with ordinary market forces. Instead it recommended a strongly interventionist strategy designed to overcome what it saw as massive market failures inherent in Ireland's situation as a country engaged in late industrialisation under free trade conditions. But the Report did not give any similar endorsement to special measures to develop small firms.

In fact, the Telesis Report is rather ambivalent in its approach to small firm development. While the whole thrust of the Report is towards "the building of fewer larger companies with strong internal capability", it goes on to say that this is not regarded as a substitute for the development of small industries but rather "as a supplement in those cases where the business opportunity is large and the chances for success are great" (p. 234). Though the Report is strongly critical of the extent of the "hand-holding" activities by the development agencies, it, nevertheless, suggests that strong regional offices could maintain some of the intensive hand-holding functions of the Shannon Free Airport Development Company (SFADCo) — functions which went beyond most of the other development agencies. It is clear, however, that the Report favours the handling of small industry projects "as part of an integrated indigenous development charter" to build structurally strong firms and to foster linkages, rather than dealing with them in any special way or by any special agency or division.

A case can indeed be articulated against a wide-ranging policy directed specifically to small industry. The bulk of this sector is engaged in non-traded activities and its potential for expansion is limited by the size of the domestic market. Given an absence of overseas competition, it can be argued that new firm formation and expansion of existing firms would proceed anyway without State incentives. These incentives may operate to displace other firms which have not received them, and may have little overall effect on employment if the total market is not enlarged either by export growth or import substitution. The high degree of flux in the small firm sector, the great diversity of firms, and the large numbers involved, could make for a proliferation of interventions that in the aggregate would be costly in relation to the overall net benefits.

The Case for Special Attention to Small Industry

While the foregoing considerations are not without force, nevertheless for the following reasons we believe that the approach implied in the Telesis Report does not give adequate weight or attention to the potential contribution of small industry.

First, it should be pointed out that it will take a lot of time and effort to operationalise any new strategy, and it would seem only sensible to keep in place the main elements of the present strategy until new approaches have been worked out.

Second, the need for jobs is very pressing in the next 5–10 years, and we have shown that the SIP has been a significant contributor to employment at a relatively low capital cost. No doubt some of the projects were not ideal from a long-term point of view and will not remain viable. But it remains to be demonstrated in practice that an alternative approach would yield better results for the same costs, and until this has been shown at least through pilot experiments, it would be risky to forgo the achievements of the present approach. Indeed, we have already suggested in the previous section that even in the long-term small industry should be looked to as an important source of manufacturing jobs. In so far as lower wages are important to employment creation, small firms provide a degree of flexibility that may be unattainable in larger enterprises. Moreover, it was suggested in Chapter 4 that small firms may provide employment for those who would find difficulty in adapting to the needs of larger-scale industry.

Third, it is clear that there are many opportunities potentially available to sub-suppliers, given that only a small proportion of the sub-supply needs of the new large firms have been met domestically. These opportunities, however, will continue to be unrealised unless the skill levels sub-supplying small firms can be upgraded, and there is no evidence that this will happen quickly through the unaided operation of market forces.

Fourth, the Telesis Report pays no attention to the important issue of regional policy. We have shown that small firm projects can play an important role as a component of regional policy because of their adaptability to locations where large industry would not be viable.[46] And as discussed in Chapter 6, this role is now recognised in the regional policies of most governments.

Fifth, Ireland already has a sizeable number of small firms, employing over one-fifth of the total engaged in manufacturing and more than one-third of those engaged in indigenous firms. The majority of the small firms will never become large, but their survival is not a matter of indifference from the standpoint of employment preservation. Clearly in view of the high degree of flux demonstrated in the small firm sector in all countries, no policy can or should aim at the survival of all of them. Equally, however, the more of them that can be induced to raise their efficiency, the better their chances of surviving and expanding and the greater their contribution to the whole process of industrialisation. This task often poses quite different issues for those involved in dealing with large firms.

Sixth, there is the case of the minority of small firms that can become large. Given the poor record to date of the generality of medium-to-large indigenous firms, it would be unwise to look to this section alone as the base for building a selected number of large strong firms. Moreover, the absence in Ireland of a significant "spin-off" process, characteristic of more developed industrial structures, has been noted by several scholars (Cogan and Onyenadum, 1981 and Murray, 1983). To enlarge the pool of potentially dynamic enterprise, therefore, it will be necessary also to look to the promotion of new first-time enterprise and the further development of the more promising of the existing small firms that have shown a potential for growth. Given the great uncertainty that inevitably attaches to enterprise development, the small industry sector provides a relatively cheap experimental setting.

Finally, the outlook and the instruments needed to nurture new or small enterprise are very different, at least in the initial stages, from those required for the

[46]New foreign firms, also have tended to serve regional policy purposes better than large indigenous firms, in that they have often been more willing and able to accept quite peripheral locations.

enlargement of firms that have already acquired a certain scale and experience. For example, the Telesis Report regards it as vital that the strong firms be encouraged to provide their own capabilities in research, marketing, etc., something that would neither be feasible nor economic in many new or small firms until they had reached a reasonable scale. More generally the capacity of small firms to receive help is not the same as large. Moreover, the energies of the development agency staff responsible for the large companies would be dissipated if they were also responsible for building from scratch the pool from which they were drawing. Given the very substantial resources that will be committed to each of the firms selected for enlargement, it would seem best also that the selection be done by a separate set of personnel from those responsible for bringing the firm to that level, and who could be quite understandably less detached in their assessment.

For the foregoing reason, therefore, we believe that there is need for specific attention to the development of small industry. Clearly, however, the small firm strategy, either in its conception or its administration, should not be isolated from the overall industrial strategy. For that reason, central responsibility for implementing the strategy should rest with a special division within the agency primarily responsible for implementing the overall strategy. We turn now to what the small industry strategy for the future should be.

New Strategic Directions

While the Government White Paper (1984) on Industrial Policy did not spell out its reasons for special attention to small industry, it, nevertheless, envisages the continuation of the Small Industry Programme, subject to certain modifications. The programme is to be put on a regional basis over a two-year period, with all decisions on projects taken locally. The IDA is to be given responsibility for co-ordinating the full range of state services to small industry (except in the area covered by SFADCo and Údarás na Gaeltachta). Thus a "one-stop shop" will be provided in each region for State advisory, training and support services for small firms.

In accordance with the approach to manufacturing as a whole, there is to be greater selectivity in grants to small industry. These will be confined to firms directly exporting, or supplying high skill, high technology goods to larger exporting firms, or displacing imports in activities where imports constitute at least 25 per cent of home sales. In practice, however, the third of these criteria would allow in most of the activities that are now eligible, though it is stated in the White Paper that in future "the onus of proof for import displacement will be on the firm seeking grant assistance" (p. 37). It is not made clear whether the grant rates would continue to be higher for small than for large industry. There is, however, a clear desire to secure greater equity involvement either by the small firms themselves or by private investors in financing small firms — with a target of a minimum ratio of equity to total project cost of 40 per cent. Moreover banks and lending agencies are to be given a greater incentive to vet and monitor projects, since in future the IDA will limit loan guarantees to 80 per cent. The White Paper also announced a number of initiatives in regard to the development of linkages, sub-supply and co-operative trading, which will be tested first on a pilot basis.

In the light of Irish circumstances, it seems to us that there is a good case for a more explicit two-tiered approach to small industry. The *first* tier would be concerned with raising efficiency in a wide range of small firms in the interests of employment creation, realisation of sub-supply opportunities, and regional policy — even though most of them will never grow big. The *second* tier would seek to identify and develop a selected number of new and existing small industries to the point where they could be handed over as possible candidates for further enlargement by the division responsible for building structurally strong companies. We envisage that rather different approaches and policy instruments should apply to these two tiers, which we now discuss.[47]

3 *Policy Instruments*

The first tier would operate much like the present SIP, with some reforms. It could best be handled on a regional basis, since the level and type of assistance can be standardised sufficiently to enable effective delegation without undue anomalies arising. The present administrative upper limit of 50 persons engaged would continue to apply to firms dealt with in this way. While being selective, it would be only moderately so and the objective would be to keep the administration of the scheme as simple as possible. Expansion grants would be related to a more realistic view of prospective employment increases, and set at a limit per job by reference to the growing number of other state job creation schemes. There is a case for allowing flexibility in the uses to which the grant is put, rather than relating it primarily to fixed assets.

There is also a case, as Telesis has suggested, for helping firms secure more of a necessary finance by way of loan guarantees and less by way of non-repayable grants. Loan guarantees might extend to working capital, which is generally not grant-aided at present, and which can be a serious barrier to the entry and expansion of small firms. The repayment schedule should recognise the distorting impact in inflationary conditions of high nominal interest rates, as well as the

[47]The Oireachtas Joint Committee on Small Businesses (1984) Report drew a distinction analogous to that used here "between small firms which remain small but healthy, and small firms with the potential to be large" (p. 2). The Committee, however, did not develop the implications of the distinction for policy purposes.

fact that the break-even point, even for viable projects, is generally not reached for a few years. This approach could add to the risk that the guarantee would be called upon, but without such an approach, expansion could be arrested in many cases that would prove viable. Carroll (1985) reports that in the PIMS data on new business units started by large corporations, the average unit took six to eight years to break even, and longer still to earn a satisfactory return. While, clearly, independent units could never survive so long, the figures provide a salutary reminder that instant profitability is rarely attainable, and point to the need for financial arrangements that will give promising firms a reasonable time to prove themselves.

Obviously, there would have to be centralised directives in regard to sectors where there is already over-capacity. Provided the cost per job to the State were kept at a moderate level, there should be tolerance of a reasonable failure rate. In a sense, this tier of the small industry programme could be looked on as a satisfactory form of "third sector" activity, which would provide employment in an effective and reasonably low cost fashion. But it would also have the merit of giving an opportunity for risky ventures or untried entrepreneurs to prove themselves, and perhaps qualify at a later stage for inclusion in the second tier. This would go some way to meeting the key difficulty identified by Rothery (1977), namely, how to "aid the true prospective successful entrepreneur at the point where he needs it most — when he is making the vital step from employee to employer".

This first tier of the programme would also be concerned to raise efficiency in existing small firms that would not qualify for an expansion grant — because of the poor growth prospects of the activity, the unwillingness or incapacity of the management to expand, or the danger of displacing existing activity. The regional staff would be expected to maintain contact with all small firms in their area, and keep them informed of the advisory services on which they could draw. As proposed in the Government White Paper (1984) these regional offices should have responsibility for co-ordinating the different public agencies offering services, and where possible having these at a common location. Consideration might also be given to attaching to each regional office a small corps of field advisers. These would provide an extension service akin to that offered to Irish farmers by the agricultural institutions. Rather than waiting for firms to seek information and advice, the advisers would seek out the firms before their problems had reached a crisis. This approach would place more emphasis on prevention through anticipation rather than cure. Such an approach has been operated by SFADCo in the mid-west region since 1978, and an evaluation of the costs and benefits of its operation would provide guidance on whether it should be extended nationally.

The prime consideration in selecting firms for the second tier, whether they were first-time or established small firms, would be that they had solid growth prospects and a willingness on the part of their proprietors to co-operate in the steps needed to realise that growth potential quickly. The objective would be to lift the firms speedily into the middle-size range. Afterwards, a proportion of them would, helpfully, qualify for selection for further development into the large, structurally-strong companies, favoured by Telesis. An important criterion in choosing firms for the second tier would be the capability of the management team, with suitable help and advice, to bring about and maintain enlargement. Obviously this cannot be assessed definitively at the start, but what can be observed is the ability to learn rapidly and adapt — qualities emphasised by Murray (1984). There would be a presumption in favour of building on selected firms at an early stage of the life cycle of an industry — where product qualify and performance are key competitive elements — though probably more so at what O'Brien (1985) calls the early "growth phase" rather than the "emergent phase". It should be borne in mind, however, that cash flow pressures in developing innovative products can be eased by manufacturing and selling a stable technology product in the short term (Lynam, 1982).

Some track record in exporting would also be desirable: certainly the capacity to penetrate export markets at an earlier stage than in larger countries is essential in Ireland to the achievement of rapid firm growth. Long-run growth and viability are more likely to be assured, however, by concentrating on achieving high shares in segmented markets rather than building many diffuse export positions with weak market shares. There is evidence that a record of international work experience is likely to prove helpful to new founders (Murray, 1983).

The firms in the second tier should continue to be handled by the Small Industries Division until they had reached a level of, say, 100 employees, and there would be a good case for merging the Enterprise Development Programme into the second tier. Employment alone need not be the chief consideration in relation to second tier firms if they possess characteristics of importance to other sections of Irish industry. The Confederation of Irish Industry (1983) Report advocates the general use of 100 employees or less as the measuring rod for all small industry programmes. They argue that the often rapid increase in employment up to a limit of 40–50 employees reflects to some extent the more favourable incentive package currently open to firms with less than 50 employees. It is more likely, however, that the blockage is mainly due to other factors such as the critical change in management structure which a firm must undergo if it is to progress much beyond a certain threshold. We believe that because of the small number of firms capable of significant further enlargement, and the increasing commitment of resources involved, a high degree of selectivity must apply beyond the 50 employee stage, and that the arrangements proposed above are more appropriate than a general extension of small industry aids to firms with up to 100 employees.

The question would also arise as to whether the second tier should be handled regionally or nationally. The argument in favour of regional control is that a continuing relationship between particular staff and a growing company is very important, so that it would be more effective for the same staff to follow through with a particular firm from the first to the second tier. While this is a cogent argument, we, nevertheless, feel that the balance of advantage would lie in handling the second tier nationally. The whole idea of the second tier is to limit it to firms with rapid growth prospects. The firms involved will quickly have to be planning to penetrate international markets and the kind of services they need are best handled nationally. Moreover, there is a danger that if the programme were operated regionally, there would be strong pressures to extend "second-tier" treatment to firms that really were not qualified. This approach does not, of course, rule out a high degree of regional involvement with second-tier firms. On the contrary, most of the firms would be proposed for inclusion by the regional officers, who would be expected to put forward for consideration the more promising prospects among the existing small firms in their areas, or new applicants for the first tier who might better be included from the start in the second tier. To encourage the regional offices to co-operate fully in this process, it should be made clear that their performance would be judged, *inter alia*, on a number of *qualifying* applicants proposed by them relative to the extent of their region.

While the incentives offered to second-tier firms would be more extensive and generous, they would also be accompanied by a tougher attitude towards performance. If small firms are to grow, they have to be prepared to dilute their equity and develop a management structure appropriate to a larger scale. Unwillingness by the proprietors to undertake the necessary measures should be taken as evidence that the firm was not seriously committed to the steps needed to sustain expansion, and it should be referred back to the first tier. The development authority would need to be given a high degree of flexibility in the range and extent of incentives offered. The scale of these incentives should be in line with those to be used in developing Telesis-type large companies, making due allowance for differences in size — though as suggested below they would not necessarily be in the same form. Clearly the development of these firms would require a close and continuing relationship with the development authority. The question also arises as to whether the State, given the generosity of the incentives, should not provide for at least partial repayment of grants by successful candidates or, alternatively, through sharing in profits by giving some of the funding in the form of equity participation. While there is a case for this, it is an issue that should be settled on the same basis for second-tier small firms as for large firms. It should also be noted that firms in the second tier are precisely the type that would interest private venture capital institutions, and every effort should be made to develop the range of such facilities.[48] One of the advantages of the wider development of equity participation both by state and private interests lies, not only in the injection of finance provided, but also in the pressure for expansion that such outside interests would be likely to exert on a proprietor who might otherwise rest content at a modest scale.

The form of incentives for second-tier small firms would necessarily differ from those appropriate to large-scale companies. Marketing is often an area of particular neglect in small firms, a common error being to assume that because the product is good the market will know all about it. But incentives suited to large firms, which are on a scale sufficient to support an extensive marketing division, may be of limited effectiveness in regard to many small firms, which cannot sustain an effective independent marketing organisation. In this connection, there would be attractions in emulating the Japanese experience of inter-firm co-operation in functions that cannot be done satisfactorily by a single firm. It would be in keeping with the objectives of the second-tier programme to insist that firms be prepared to enter into such co-operative arrangements with other firms — and if need be to tailor the scale of incentives accordingly.

Indeed, for both small and large firms, serious examination should be made of the possibility of encouraging industrial marketing agencies exploited with such success in Japan, and which handle a high proportion of Japanese trade. The basic advantages would be in supporting selected firms to reach a viable scale of operation, in sharing the advantages of economies of scale in selling and purchasing, and in the spreading of risks. In this regard it would be useful to investigate the reasons for the failure of the Irish National Trading Corporation, as well as studying the essential pre-conditions for success underlying the Japanese experience to see if these can be remodelled to fit Irish conditions.

Another initiative exploited with particular success in Japan has been that of sub-contracting of input requirements by the large firms. Large manufacturing firms claim that Irish subsuppliers cannot meet their requirements as regards quality, standards and delivery. Clearly if this situation is to be improved there is need to establish closer contacts between large firms and potential sub-suppliers. Large firms might be encouraged to "adopt" one or more small firms, with which they would maintain a long-term relationship, including perhaps the secondment of personnel with expertise that needed to be fostered in the small firm. In developing this further in Ireland, there would be need for close liaison

[48]The consultants' report in NESC (1984) stated that venture capital companies were eager to invest in firms with a clearly-established capability for further high growth, but could not in practice find enough candidates. The consultants also recommended that the second and further stages of grant-aid should be conditional on a matching injection of private equity — either in the form of cash investment or realisable retained earnings.

between the staff of the development agency responsible for large firms and those dealing with second-tier small firms.[49] It is in this nexus above all that the potential complementarity between large and small firms can be realised, with advantages to both. Moreover, a wealth-creating approach in the large company would be more widely acceptable if it were transparent from the start that it was associated with output and employment benefits to other firms.

Similar links might also be considered in relation to the purchases of public bodies, under which the buying agencies would use their purchasing power as a lever to develop these firms. The implementation of a coherent government purchasing policy would need to be done in conjunction with the development authority and should be aimed at encouraging innovation and the raising of standards rather than as a crude protectionist device. The objective would be to give a start to those firms capable of meeting high standards of quality and design, and on the basis of which they would have good prospects of eventually competing successfully in export markets. Any price advantage accorded to such firms should therefore be phased out.

Three steps mooted in the Government White Paper (1984) could help to give effect to the foregoing ideas. One is the proposal that the IDA, CTT and the National Development Corporation should experiment on a pilot basis with the concept of the "development company". Such companies would engage in marketing, and provide finance and technical advice, but would sub-contract production to smaller firms. The second proposal is the National Linkage Programme, to be concentrated first in the electronics sector, which is designed to ensure that a much higher proportion of the raw materials and component inputs of large export firms are supplied domestically. The third proposal is the encouragement of "development contracts" by large foreign and indigenous firms, by supermarket chains and by State industries.[50] The suggestion is that the initial development costs of the supplier would be shared by the purchaser on a trial basis. These proposals are unlikely to become effective without a highly active role by the development agencies, along the general lines recommended in the Telesis report.

As in the case of marketing, grants to small firms to provide their own R & D capability may not be effective in many cases because of small scale of operation and absence of technical personnel. Small firms generally also need access to advisory services and laboratories with facilities for design experimentation and

[49]The IDA has in fact been given responsibility for the National Linkages Programme mentioned in the Government White Paper (1984), and will initially concentrate on the electronics sector.

[50]The possibility of government departments and their product and service processing agencies fostering development contracts, however, is not included in the White Paper, even though the NBST (1983a) advocated co-ordinated procurement of public goods and services for all government departments through the Department of Industry, Trade, Commerce and Tourism.

prototype development. In meeting these needs, it is far from evident that the variety of public agencies with such functions are able to plan and operate in an integrated manner at present. Competition between R & D centres can perhaps stimulate effort, but a certain scale is required before substantial applied work can be undertaken in science and technology. Regarding the relationship between industry and the higher education centres, the NBST has sought to promote closer contacts between the two in industrial R & D. Additional work needs to be done, however, to dismantle the barriers which inhibit closer collaboration between colleges and industry. Even then, however, the evidence from other countries cited in Conniffe and Kennedy (1984) suggests that the encouragement of industrial innovation via the higher education system is often problematic, and must be augmented through other channels.

4 *Concluding Comments*

The development of small firms poses a variety of problems, many of them quite distinct from those involved in building large firms. Indeed as Bannock (1981, p. 104) aptly notes, "To treat small firms in the same way as large is usually, in fact, to discriminate against them". While policy for small industry should be determined within the framework of overall industrial policy, there is a good case in our view for continued specialised attention to small firms within the major development agency, the IDA, and also perhaps in some of the other development agencies — subject to adequate co-ordination of services offered by different agencies. The major benefits to be expected for this specialised attention are two-fold: first, the enhancement of the contribution of manufacturing to employment creation in a wide range of small firms, and, second, the growth of some of them into medium-sized firms, with potential for still further enlargement in a small minority of cases.

Nevertheless, there is a danger in devoting special attention to small firms that the great diversity in their experience and in the problems facing them, could lead to a continuous accretion of services. Thus a service introduced to meet problems particular to one class of firms may be demanded even by those with less justification for receiving that benefit. We believe that the two-tier strategy advocated in this chapter would forestall much of this pressure.

Some further safeguards can also be suggested. First, all programmes should try to establish clear objectives and criteria, showing how these are related to the overall strategy. Second, to facilitate this process and to ensure that programmes are as effective as possible, there should be pilot experimentation with any new service or incentive before it is introduced generally, or scrapped. The pilot-testing of schemes by SFADCo in the mid-west region has set the headline in that regard. Third, the progress of each of the programmes should be monitored by the parent government department and not just by the agency responsible for the programme. Fourth, the aggregate of services and incentives offered by the various agencies should be subject to independent overall evaluation at periodic intervals not only to ensure adequate co-ordination, but also to re-assess the scale of subsidisation in relation to benefits.

Finally, while individual programmes can be modified or dropped without undue disruption, the broad framework of any strategy must remain in place for a considerable time. This is necessary if clear signals are to be given to firms and if the development agencies are to market and implement the strategy satisfactorily. That being so, it is important, despite the present pressing economic needs, that adequate thought and attention be given to considering the longer-term implications of alternative approaches. In short, we believe it would be a mistake to rush into a new strategy dominated by immediate economic concerns. Rather the objective should be to seek to set in place in the next few years the framework of industrial strategy that will be needed to carry to the turn of this century. This study is offered as a contribution to that process.

REFERENCES

ABERNATHY, W. J. and J. M. UTTERBACK, 1976. *Innovation and the Evolution of Technology in the Firm*, Harvard: Business School.

AHMED, M., 1976. "Size and Efficiency: The Case of the Irish Plastics Industry", *The Economic and Social Review*, Vol. 8, No. 1, October, 1-21.

ALLEN, T. J., 1979. "Transferring Technology to the Firm: A Study of the Diffusion of Technology in Irish Manufacturing Industry", *Journal of Irish Business and Administrative Research*, Vol. 1, No. 2, October.

ALLEN, T. J. and V. REILLY, 1973. "Getting the Word About — Technology Transfer to Irish Industry", Working Paper No. 650/73, Massachusetts Institute of Technology.

ANDERSON, T. A., 1967. "The Effects of Size on Profits in Manufacturing Industries" in I. Pfeffer (ed.) *The Financing of Small Business*, New York: Macmillan.

ARMINGTON, C. and M. ODLE, 1982. "Small Business — How Many Jobs?" *The Brookings Review*, Winter.

BANNOCK, G., 1976. *The Smaller Business in Britain and Germany*, London: Wilton House.

BANNOCK, G. 1981. *The Economics of Small Firms: Return from the Wilderness*, Oxford: Basil Blackwell.

BATES, J., 1971. *The Financing of Small Business*, Second Edition, London: Sweet and Maxwell.

BAUMOL, W. J., 1968. "Entrepreneurship in Economic Theory", *American Economic Review*, Vol. LVIII, No. 2, May, 64-71.

BECKLER, D., 1980. "Small and Medium-Size Firms in the OECD Countries — Trends and Policy Issues", paper given at International Symposium on *Industrial Policies for the 1980s*, Madrid, May 5-9.

BEESLEY, M. and P. WILSON, 1982. "Government Aid to the Small Firm Since Bolton" in J. Stanworth, G. Westrip, D. Watkins and J. Lewis (eds.) *Perspectives on a Decade of Small Businesses Research: Bolton Ten Years On*, Aldershot: Gower Publishing Company.

BINKS, M., 1979. "Finance for Expansion in the Small Firm", *Lloyds Bank Review*, No. 134, October.

BINKS, M. and J. COYNE, 1983. *The Birth of Enterprise: An Analytical and Empirical Study of the Growth of Small Firms*, Hobart Paper No. 98, London: Institute of Economic Affairs.

BIRCH, D. L., 1979. *The Job Generation Process*, Massachusetts Institute of Technology Report.

BOARD OF TRADE, 1967. *Economic Trends*, No. 168, November.

BOLLARD, A., 1983. "Technology, Economic Change and Small Firms", *Lloyds Bank Review*, No. 147, January.

BOLLARD, A. 1984. "Small Firms and Industrial Change in New Zealand", *Quarterly Predictions*, New Zealand Institute of Economic Research, June.

BOLTON, J. E., 1971. (Chairman) *Small Firms: Report of the Committee of Inquiry on Small Firms*, London: HMSO.

BOSWELL, J., 1972. *The Rise and Decline of Small Firms*, London: Allen and Unwin.

BRADLEY, M. F., 1983. "A Public Policy for International Marketing", *Journal of Irish Business and Administrative Research*, Vol. 5, No. 2, October.

BUREAU OF INDUSTRY ECONOMICS, 1981. *Finance for Small Business Growth and Development*, Research Report No. 10, Australian Government Publishing Service, Canberra.

CALLINAN, B., 1984. "The Work of Shannon Free Airport Development Company", *Administration*, Vol. 32, No. 3.

CARROLL, C., 1985. *Building Ireland's Business: Perspectives from PIMS*, Dublin: Irish Management Institute.

CARSON, D. and I. O. SCOTT, 1973. "Differential Effects of Monetary Policy on Small Business" in D. Carson (Ed.) *The Vital Majority: Small Business in the American Economy*, Washington DC: Government Printing Office.

CENTRAL STATISTICS OFFICE, 1973. *Analysis of the Census of Industrial Production 1968*, Supplement to *Irish Statistical Bulletin*, March.

CENTRAL STATISTICS OFFICE, 1984. *Census of Industrial Production 1979*, Dublin: Stationery Office.

CENTRAL STATISTICS OFFICE, 1985. *Population and Labour Force Projections 1986–1991*, Dublin: Stationery office, April.

CHAMBERLIN, E. H., 1956. *The Theory of Monopolistic Competition*, 7th Edition, Cambridge: Harvard University Press.

CHANDLER, A. E., 1962. *Strategy and Structure: Chapters in the History of Industrial Enterprise*, Cambridge, Mass.: MIT Press.

COGAN, D. and B. McGOVERN, 1984. "Promoting New Technology-based Companies: The Role of the IDA Enterprise Development Programme", *Journal of Irish Business and Administrative Research*, Vol. 3, No. 2, October.

COGAN, D. and E. ONYENADUM, 1981. "Spin-off Companies in the Irish Electronics Industry", *Journal of Irish Business and Administrative Research*, Vol. 3, No. 2, October.

COMMISSION OF THE EUROPEAN COMMUNITIES, 1984. *Operations of the European Community Concerning Small and Medium-sized Enterprises and Craft Industry: Practical Handbook*, Revised version of 20 July.

CONFEDERATION OF IRISH INDUSTRY, 1978. *Small Firms Report*, Dublin: CII Small Firms Association.

CONFEDERATION OF IRISH INDUSTRY, 1983. *Development Proposals for Small Firms 1983*, Dublin: CII Small Firms Association.

CONNIFFE, D. and K. A. KENNEDY, 1984. (Editors) *Employment and Unemployment Policy for Ireland*, Dublin: The Economic and Social Research Institute.

COOPER, C. and N. WHELAN, 1973. *Science, Technology and Industry in Ireland*, Report to the National Science Council, Dublin: Stationery Office.

CORAS TRACHTALA, 1980. *29th Annual Report and Accounts*, Dublin: CTT.

CURRAN J. and J. STANWORTH, 1979. "Worker Involvement and Social Relations in the Small Firm", *The Sociological Review*, Vol. 27, No. 2, May.

DEPARTMENT OF FINANCE, 1983. *Economic Review and Outlook*, Dublin: Stationery Office.

ECONOMIC COMMISSION FOR LATIN AMERICA, 1981. *Statistical Yearbook for Latin America*, United Nations Commission for Latin America, Santiago, Chile.

ECONOMIC PLANNING BOARD, JAPAN, 1957. "Statistical Analysis of Medium and Small Enterprises in Japan", *Asian Affairs*, Vol. II, No. 2, June, pp. 195–217.

FARLEY, N. J. J., 1975. "Determinants of Establishment Size in irish Manufacturing Industries: Some Notes on the Irish Case 1931-1972", *The Economic and Social Review*, Vol. 6, No. 2, January, pp. 187–213.

FOGARTY, M. P., 1973. *Irish Entrepreneurs Speak for Themselves*, Dublin: The Economic and Social Research Institute, Broadsheet No. 8, December.

FOLEY, A., 1977. "Productivity and Size of Plant in Irish Manufacturing", M.A. Dissertation, University College, Dublin.

FOTHERGILL, S. and G. GUDGIN, 1979. *The Job Generation Process in Britain*, Centre for Environmental Studies Research Series No. 32, November.

FRANSMAN, M., 1984. "Explaining the Success of the Asian NICs: Incentives and Technology", IDS Sussex *Bulletin*, Vol. 15, No. 12.

FREEMAN, C., 1971. *The Role of Small Firms in the UK since 1945*, Bolton Committee of Inquiry on Small Firms, Research Report No. 6, London: HMSO.

FREEMAN, C., 1982. *The Economics of Industrial Innovation*, 2nd Edition, London: Frances Pinter.

GARVIN, W. J., 1968. "An Economic Evaluation of the Business Loan Program of the Small Business Administration", paper presented to the Institute of Management Science, September.

GIBRAT, R., 1931. *Les Inegalites Economiques*, Paris.

GILDER, G., 1981. "The Role of the Entrepreneur", *Dialogue*, No. 53 (reprinted from *The American Spectator*).

GONENC, A., 1984. "Venture Capital: Key Ingredient in High-Tech Start-ups", *OECD Observer*, No. 131, November.

GORMAN, L. and E. MOLLOY, 1972. *People, Jobs and Organisations*, Dublin: Irish Management Institute.

GORMAN, L., R. HANDY, T. MOYNIHAN and R. MURPHY, 1974. *Managers in Ireland*, Dublin: Irish Management Institute.

GORMAN, L., G. HYNES, J. McCONNELL and T. MOYNIHAN, 1975. *Irish Industry: How It Is Managed*, Dublin: Irish Management Institute.

GOVERNMENT WHITE PAPER, 1984. *Industrial Policy*, Dublin: Stationery Office.

HALL, M. and L. WEISS, 1967. "Firm Size and Profitability", *Review of Economics and Statistics*, Vol. XLIX, No. 3, August.

HAMBERG, D., 1964. "Size of Firm, Oligopoly and Research: The Evidence", *Canadian Journal of Economics and Political Science*, Vol. 30, No. 1.

HART, P. and S. PRAIS, 1956. "The Analysis of Business Concentration: A Statistical Approach", *Journal of the Royal Statistical Society*, Part II.

HIGGINS, T., 1977. *Research, Planning and Innovation*, Dublin: Stationery Office.

HUTCHINSON, P. J., 1978. "Financial Assistance to Small Firms: The American Experience", *National Westminster Quarterly Review*, November.

HYMER, S. and P. PASHIGIAN, 1962. "Firm Size and the Rate of Growth", *Journal of Political Economy*, Vol. 70, December.

HYMER, S. and P. PASHIGIAN, 1964. "Reply", *Journal of Political Economy*, Vol. 72, February.

IJIRI, Y. and H. SIMON, 1964. "Business Firm Growth and Size", *American Economic Review*, March.

IJIRI, Y. and H. SIMON, 1967. "A Model of Business Firm Growth", *Econometrica*, April.

IJIRI, Y. and H. SIMON, 1974. "Interpretations of Departures from the Pareto Curve Firm Size Distributions", *Journal of Political Economy*, Vol. 82, No. 2.

INTERNATIONAL LABOUR ORGANISATION, 1956. "The Size of Industrial Establishments", *International Labour Review*, June.

JACKSON, J., 1979. *Small Independent Enterprises and their Environment: Factors Affecting the Maintenance and Growth of Employment among Small Independent Enterprises*, Main Report for Ireland to the Commission of the European Communities.

JEWKES, J., 1952. "The Size of the Factory", *Economic Journal*, June, pp. 237–251.

JEWKES, J., D. SAWERS and R. STILLERMAN, 1969. *The Sources of Invention*, Macmillan.

KAMIEN, M. I. and N. L. SCHWARTZ, 1975. "Market Structure and Innovation: A Survey", *Journal of Economic Literature*, Vol. 23, No. 1.

KENNEDY, K. A., 1971. *Productivity and Industrial Growth: The Irish Experience*, Oxford: Clarendon Press.

KENNEDY, K. A., 1982. "The Design of a New Industrial Strategy", paper read to a Symposium on Industrial Development, *Journal of the Statistical and Social Inquiry Society of Ireland*, Vol. XXIV, Part V, pp. 34–41.

KIERAN, J., 1975. *Technology Transfer*, Dublin: Stationery Office.

KLEINMAN, E., 1971. "Wages and Plant Size: A Spillover Effect?" *Industrial and Labour Relations Review*, Vol. 24, No. 2, January.

LEFF, N. H., 1979. "Entrepreneurship and Economic Development: The Problem Revisited", *The Journal of Economic Literature*, Vol. 17, No. 1, March, pp. 46–64.

LEE, J., 1983. "Motivation: A Historian's Point of View", in *Ireland in the Year 2000 – Towards a National Strategy: Issues and Perspectives*, Proceedings of a Colloquy, Dublin: An Foras Forbartha.

LEIBENSTEIN, H., 1968. "Entrepreneurship and Development", *American Economic Review*, Vol. 58, No. 2, May, pp. 72–83.

LINEHAN, T. P., 1962. "The Structure of Irish Industry", *Journal of the Statistical and Social Inquiry Society of Ireland*, Vol. XX, Part V, pp. 220–246.

LITTLE, A. D., 1977. *New Technology Based Firms in the United Kingdom and the Federal Republic of Germany*, a report prepared for the Anglo-German Foundation for the Study of Industrial Society.

LYNAM, H., 1982. "Technology-based Small Manufacturing Enterprises: A Personal View", *Journal of Irish Business and Administrative Research*, Vol. 4, No. 1, April.

McMILLAN COMMITTEE, 1931. *Report of the Committee on Finance and Industry*, Cmd 3897, London: HMSO.

McALEESE, D., 1977. *A Profile of Grant-aided Industry in Ireland*, Dublin: Industrial Development Authority.

McALEESE, D., 1983. "Is there a Want of Capital in Ireland?" lecture to mark the Golden Jubilee of the Industrial Credit Company, Dublin, October.

MANSFIELD, E., 1962. "Entry, Innovation and the Growth of Firms", *American Economic Review*, Vol. 52, No. 5, December.

MARGINSON, P. M., 1984. "The Distinctive Effects of Plant and Company Size on Workplace Industrial Relations", *British Journal of Industrial Relations*, Vol. 22, No. 1, March.

MERRETT CYRIAX ASSOCIATES, 1971. *Dynamics of Small Firms*, Bolton Committee of Inquiry on Small Firms, Research Report No. 12, London: HMSO.

MEYER, J. and E. KUH, 1957. *The Investment Decision*, Cambridge, Mass.: Harvard University Press.

MINISTRY OF FOREIGN AFFAIRS, JAPAN, 1982. *Small Businesses in Japan – Underpinnings of Economic Vitality*, Tokyo.

MINISTRY OF INTERNATIONAL TRADE AND INDUSTRY, JAPAN, 1978. *White Paper on Small and Medium Enterprises*, Tokyo.

MURRAY, J. A., 1981. "In Search of Entrepreneurship", *Journal of Irish Business and Administrative Research*, Vol. 3, No. 2, October.

MURRAY, J. A., 1983. "Entrepreneurship and Innovation: Some Research Findings", *Journal of Irish Business and Administrative Research*, Vol. 5, No. 1, April.

MURRAY, J. A., 1984. "Strategic Behaviour in the New Venture", *Journal of Irish Business and Administrative Research*, Vol. 6, No. 1, April.

NATIONAL BOARD FOR SCIENCE AND TECHNOLOGY, 1981. *Irish Science and Technology Statistics*, Dublin.

NATIONAL BOARD FOR SCIENCE AND TECHNOLOGY, 1982. *Innovation in Small Manufacturing Firms: An Irish Research Study*, Dublin.

NATIONAL BOARD FOR SCIENCE AND TECHNOLOGY, 1983a. *Science Budget 1983*, Dublin.

NATIONAL BOARD FOR SCIENCE AND TECHNOLOGY, 1983b. *Irish Science and Technology Statistics*, Dublin.

NATIONAL SCIENCE FOUNDATION, 1977. *Science Indicators 1976*.

NATIONAL ECONOMIC AND SOCIAL COUNCIL, 1982. *Policies for Industrial Development: Conclusions and Recommendations*, Report No. 66, Dublin.

NATIONAL ECONOMIC AND SOCIAL COUNCIL, 1983. *An Analysis of Job Losses in Irish Manufacturing Industry*, Report No. 67, Dublin.

NATIONAL ECONOMIC AND SOCIAL COUNCIL, 1984. *The Role of the Financial System in Financing the Traded Sectors*, Dublin, Report No. 76, October.

O'BRIEN, R., 1985. "Technology and Development: The Irish Electronics Industry in an International Context", in J. Fitzpatrick and J. Kelly (eds.) *Perspectives of Irish Industry*, Dublin: Irish Management Institute (forthcoming).

O'CONNOR, J. and M. LYONS, 1982. *Venture Initiation in Irish Society – A Pilot Study of the Entrepreneurial Process*, Dublin: Industrial Development Authority.

O'CONNOR, T. M., 1969. "Diversifying Employment and Social Opportunities in Local Areas: A Case Study for Ireland", paper presented to the Congress of the Industrial Federation for Housing and Planning, Dublin, May.

O'DONNELL, N. and J. MURRAY, 1983. *Development Companies: Creation, Structure, Strategy*, Dublin: The Enterprise Centre, Faculty of Commerce, University College, Dublin.

O'FARRELL, P. N., 1975. *Regional Industrial Development Trends in Ireland 1960–1973*, Dublin: Industrial Development Authority.

O'FARRELL, P. N., 1976. "An Analysis of Industrial Closures: Irish Experience 1960–1973", *Regional Studies*, Vol. 10, pp. 433–448.

O'FARRELL, P. N., 1982. "Industrial Linkages in the New Industry Sector: A Behavioural Analysis", *Journal of Irish Business and Administrative Research*, Vol. 4, No. 1, April.

O'FARRELL, P. N., 1984a. "Components of Manufacturing Employment Change in Ireland 1973–1981", *Urban Studies*, Vol. 21, 155–176.

O'FARRELL, P. N., 1984b. "New Manufacturing Plant Openings and Employment Growth", *Journal of Irish Business and Administrative Research*, Vol. 6, No. 2, October.

O'FARRELL, P. N., 1984c. "Small Manufacturing Firms in Ireland: Employment Performance and Implications", *International Small Business Journal*, Vol. 2, No. 2, Winter.

O'FARRELL, P. N., 1985a. "Employment Change in Manufacturing: The Case of Servicing Plants", *Urban Studies*, Vol. 22, pp. 57–68.

O'FARRELL, P. N., 1985b. *Entrepreneurship and New Firms*, (forthcoming).

O'FARRELL, P. N. and R. CROUCHLEY, 1983. "Industrial Closures in Ireland 1973-1981: Analysis and Implications", *Regional Studies*, Vol. 17, No. 6, pp. 411–427.

O'FARRELL, P. N. and R. CROUCHLEY, 1984. "An Industrial and Spatial Analysis of New Firm Formation in Ireland", *Regional Studies*, Vol. 18, No. 3, pp. 221–236.

O'FARRELL, P. N. and B. O'LOUGHLIN, 1980. *An Analysis of New Industry Linkages in Ireland*, Dublin: Industrial Development Authority.

O'FARRELL, P. N. and B. O'LOUGHLIN, 1981. "New Industry Input Linkages in Ireland: An Econometric Analysis", *Environmental Planning*, Vol. 13, pp. 285–308.

O'MALLEY, E., 1983. *Late Industrialisation under Outward-looking Policies: The Experience and Prospects of the Republic of Ireland*, D.Phil. Dissertation, University of Sussex.

O'REILLY, A., A. JONES and A. COLDRICK, 1980. *Keeping up to Date as a Manager*, Dublin: AnCO — The Industrial Training Authority.

ORGANISATION FOR ECONOMIC CO-OPERATION AND DEVELOPMENT, 1971a. *Problems and Policies Relating to Small and Medium-sized Enterprises*, Paris.

ORGANISATION FOR ECONOMIC CO-OPERATION AND DEVELOPMENT, 1971b. *The Industrial Policies of Fourteen Member Countries*, Paris.

ORGANISATION FOR ECONOMIC CO-OPERATION AND DEVELOPMENT, 1978. *Selected Industrial Policy Instruments: Objectives and Scope*, Paris.

ORGANISATION FOR ECONOMIC CO-OPERATION AND DEVELOPMENT, 1982. *Innovation in Small and Medium Firms*, Paris.

OSBORN, R. C., 1970. "Concentration and the Profitability of Small Manufacturing Corporations", *Quarterly Review of Economics and Business*, Vol. 10, Summer.

PAPENEK, G. F., 1982. "The Development of Entrepreneurship", *American Economic Review*, Vol., LII, No. 2, May, pp. 46–58.

PRAIS, S., 1976. *The Evolution of Giant Firms in Britain: A Study of the Growth of Concentration in Manufacturing Industry in Britain 1909–70*, Cambridge: Cambridge University Press.

PRAIS, S., 1978. "The Strike Proneness of Large Plants in Britain", *Journal of the Royal Statistical Society*, Series A, Vol. 141, Part 3.

PRATTEN, C. F., 1971. *Economies of Scale in Manufacturing Industry*, Cambridge: Cambridge University Press.

PRYOR, F. L., 1972. "The Size of Production Establishments in Manufacturing", *The Economic Journal*, Vol. 82, No. 326, June, pp. 547–566.

RADCLIFFE COMMITTEE, 1959. *Report of the Committee on the Working of the Monetary System*, London: HMSO.

ROTHWELL, R. and W. ZEGVELD, 1978. *Small and Medium Manufacturing Firms: Their Role and Problems in Innovation Government Policy in Europe, the USA, Canada, Japan and Israel*, report prepared for the Six Countries Programme on Government Policies towards Technological Innovation in Industry.

ROTHERY, B., 1977. *Men of Enterprise*, Dublin: Institute for Industrial Research and Standards.

RUANE, F. P., 1976. "Trade, Fiscal Policy and Industrialisation in the Small Open Economy: The Irish Experience", B.Phil Thesis, Nuffield College, Oxford.

RUANE, F. P., 1978. "Government Policy and Resource Allocation in Semi-developed Countries with Special Reference to the Republic of Ireland", D.Phil. Thesis, Nuffield College, Oxford.

RUANE, F. P. and P. NEARY, 1982. "Inflation and Growth" in D. McAleese and L. Ryan (eds.) *Inflation in the Irish Economy: A Contemporary Perspective*, Dublin: Helicon.

RYAN, W. J. L., 1949. "The Nature and Effects of Protectionist Policy in Ireland 1922–1939", unpublished Ph.D. thesis, Trinity College, Dublin.

SAMUELS, J. M., 1965. "Size and the Growth of Firms", *Review of Economic Studies*, Vol. 32.

SAMUELS, J. M. and W. J. SMYTH, 1968. "Profits, Variability of Profits and Firm Size", *Economica*, Vol. 35, No. 138, May.

SCHUMPETER, J. A., 1934. *The Theory of Economic Development: An Inquiry into Profits, Capital, Credit, Interest and the Business Cycle*, Cambridge, Mass.: Harvard University Press.

SCHUMPETER, J. A., 1950. *Capitalism, Socialism and Democracy*, New York: Harper, Third Edition.

SEXTON, J. J., 1982. "Sectoral Changes in the Labour Force over the Period 1961-80 with Particular Reference to Public Sector and Services Employment", *Quarterly Economic Commentary*, August.

SHINOHARA, M., 1968a. "Formation and Transition of the Dual Economy in Japan", *Hitotsubashi Journal of Economics*, Vol. 8, No. 2.

SHINOHARA, M., 1968b. "A Survey of Japanese Literature on Small Industry", in B. Hoselitz (ed.) *The Role of Small Industry in the Process of Economic Growth*, The Hague and Paris: Mouton.

SIMON, H., 1964. "Comments: Firm Size and the Rate of Growth", *Journal of Political Economy*, Vol. 72, February.

SIMON, H. and C. BONINI, 1958. "The Size Distribution of Business Firms", *American Economic Review*, September.

SINGH, A. and G. WHITTINGTON, 1968. *Growth, Profitability and Valuation*, Cambridge: Cambridge University Press.

SOETE, L. G., 1979. "Firm Size and Inventive Activity: The Evidence Reconsidered", *European Economic Review*, Vol. 12, No. 4, October.

STEWART, J., 1984. "Financial Aids to Small Business", *The Irish Banking Review*, March.

STOLL, H. R. and A. J. CURLEY, 1970. "Small Business and the New Issue Market for Equities", *Journal of Finance and Quantitative Analysis*, June.

STOREY, D. J., 1982. *Entrepreneurship and the New Firm*, London: Croom Helm.

SURVEY OF GRANT-AIDED INDUSTRY, 1967. Dublin: Stationery Office.

SWEENEY, G. P., 1981. *New Entrepreneurship and the Smaller Firm*, Dublin: Institute for Industrial Research and Standards.

TAMARI, M., 1970. *A Postal Questionnaire Survey of Small Firms: An Analysis of Financial Data*, Bolton Committee Research Report No. 16, London: HMSO.

TELESIS CONSULTANCY GROUP, 1982. *A Review of Industrial Policy*, Report No. 64, Dublin: National Economic and Social Council.

TODD, D., 1977. "Efficiency and Size: Comments and Extensions", *The Economic and Social Review*, Vol. 8, No. 4, July.

TOMLIN, B., 1966. *The Management of Irish Industry*, Dublin: Irish Management Institute.

TRIDENT MANAGEMENT CONSULTANTS, 1983. *The Performance of Udaras na Gaeltachta and of its Predecessor, Gaeltarra Eireann, in the Period 1971-1981.*

UNITED STATES COMMERCE DEPARTMENT, 1981. "US Direct Investment Overseas", *Survey of Current Business*, Washington, DC.

UTTERBACK, J. M., 1979. "The Dynamics of Product and Process Innovation in Industry", in C. T. Hill, and J. M. Utterback (eds.) *Technological Innovation for a Dynamic Economy*, New York: Pergamon Press.

UTTON, M. A., 1972. "Mergers and the Growth of Large Firms". *Oxford Bulletin of Economics and Statistics*, May.

VAUGHAN, R. N., 1980. *Measures of the Capital Stock in the Irish Manufacturing Sector 1945-1973*, Dublin: The Economic and Social Research Institute, Paper No. 103.

VERNON, R., 1966. "International Investment and International Trade in the Product Cycle", *Quarterly Journal of Economics*, May.

WALSH, B. M., 1971. "Aspects of Labour Supply and Demand with Special Reference to the Employment of Women in Ireland", *Journal of the Statistical and Social Inquiry Society of Ireland*, Vol. XXII, Part III, pp. 88-118.

WALSH, B. M., 1985. "Venture Capital in Ireland: A Review", *Journal of Irish Business and Administrative Research*, Vol. 7, No. 1, Spring.

WATANABE, S., 1970. "Entrepreneurship in Small Enterprises in Japanese Manufacturing", *International Labour Review*, Vol. 102, No. 6.

WHITE, L. J., 1982. "The Determinants of the Relative Importance of Small Business", *The Review of Economics and Statistics*, Vol. LXIV, February, pp. 42-49.

WHITE, P., 1983. "Opportunity — Ireland", Address to the Dublin Chamber of Commerce Bicentenary Conference, Dublin, September.

WILSON COMMITTEE, 1979. *Interim Report on the Financing of Small Firms*, Cmd 7503, London: HMSO.

WORLD BANK, 1978. *Employment and Development of Small Enterprises*, Washington, D.C.: International Bank for Reconstruction and Development.

Appendix

DATA SOURCES AND METHODS

Two unpublished sources of data were used in examining the Small Industries Programme: (i) the IDA *Annual Employment Survey* (for 1973 and for each subsequent year up to 1980), and (ii) an IDA *Project Information File*. In addition, data were supplied for recent years in relation to the mid-west region by SFADCo which has had responsibility since 1978 for administering the SIP in that region for small indigenous firms.

The IDA Annual Employment Survey is a record of employment in all manufacturing establishments on January 1st of each year commencing in 1973. An establishment is defined as a location in which a specific product is manufactured. Hence, a firm considered as one unit from the point of view of legal ownership may contain one or more establishments. There were 6,960 manufacturing establishments in all recorded on the Employment Survey. Of these 4,829 were recorded as being in employment in 1973, and 5,528 in 1980. Throughout this study, stated employment in any year based on IDA sources refers to employment on January 1st of that year. Some of the establishments included in the Survey are engaged in service activities, but these have been excluded throughout in this study. Where, however, a service activity is carried out at the same location as manufacturing, and more than 50 per cent of the employment is in manufacturing, the total employment may in some cases be classified in the Survey as manufacturing. The Annual Employment Survey contained a code identifying the type of IDA grant assistance, if any, for which an establishment had been approved. Deficiencies in the data that might lead to erroneous conclusions have been noted as they arise in the main body of this study.

The Project Information File contained information on 1,608 establishments which were approved for a SIP grant between 1967 and 1978. The File included data on grant approvals, proposed expenditure on fixed assets, associated job approvals, payments of grants, actual investment in fixed assets for which capital grant payments were made, as well as the dates at which approvals and payments were made.

Price Indices

Five price indices were used to deflate grant amounts and associated investment. These were:

(1) An index of prices of "Transportable Capital Goods for Use in Industry" (a Wholesale Price Index) used for deflating approved expenditure on plant and equipment (Source: *Irish Statistical Bulletin*).

(2) An index of prices of "Building and Construction" (a Wholesale Price Index, *Irish Statistical Bulletin*) used to deflate approved investment in site and buildings. This index is based on the cost of wages and raw materials in the building and construction industry.

(3) A composite capital goods price index used to deflate (i) actual investment; (ii) capital grant approvals; and (iii) capital grant payments. This index was obtained by taking the simple average of the index values for (1) and (2) for each year. The attaching of equal weights to the price indices of site and buildings and plant and equipment was based on the actual experience of 1967-76, when approved investment on these two elements of total investment were £36.7 m and £40.1 m respectively for the SIP.

Appendix Table A: *Price indices used to deflate nominal amounts of SIP grants and expenditure on fixed assets*

Year	*Plant and equipment price index* 1	*Site and buildings price index* 2	*Actual investment and capital grants price index* $3 = \frac{1}{2}(1 + 2)$	*Training grants price index* 4	*"Other incentive" grants price index* 5
1967	.523	.355	.439	.305	.441
1968	.541	.370	.456	.331	.462
1969	.567	.410	.489	.371	.496
1970	.599	.456	.528	.422	.537
1971	.639	.504	.572	.485	.584
1972	.677	.557	.617	.560	.635
1973	.723	.632	.678	.666	.707
1974	.842	.835	.839	.787	.827
1975	1.000	1.000	1.000	1.000	1.000
1976	1.189	1.161	1.150	1.186	1.144
1977	1.399	1.367	1.383	1.384	1.300
1978	1.554	1.505	1.530	1.584	1.399
1979	1.766	1.741	1.754	1.820	1.585
1980	2.014	2.095	2.055	2.155	1.873
1981	2.267	2.390	2.329	2.508	2.256
1982	2.465	2.637	2.551	2.804	2.642

Source: See text of Appendix.

(4) An index of average weekly earnings of adults in manufacturing industry used to deflate grants for the training of personnel (*Irish Statistical Bulletin*).
(5) The Consumer Price Index used to deflate "Other Incentive" grants, principally for rent subsidies (*Irish Statistical Bulletin*).

The value of each index in each year 1967–82 is given in Appendix Table A with a base year value of 100 in 1975 for all indices. The value of each index in any year is the average of quarterly observations for the same year. Grant amounts were deflated by the appropriate price indices in each case. Hence, in the case of total grant payments, capital grant amounts were deflated by the appropriate index for capital; training grants by the index of adult earnings in manufacturing; and other incentive grants by the consumer price index. The resulting real values were aggregated to arrive at the constant price value of total grant payments. Likewise, in the case of eligible fixed assets associated with grant approvals, the constituent elements (plant and equipment, and site and buildings) were each deflated by the appropriate price index. In the case of actual investment, there was an absence of reliable data on its components. Consequently, actual investment was deflated by the composite capital goods index. All the data in constant prices presented in Chapter 9 have been shown at 1982 levels, the most recent year to which the data relate.

Books:

Economic Growth in Ireland: The Experience Since 1947
Kieran A. Kennedy and Brendan Dowling

Irish Economic Policy: A Review of Major Issues
Staff Members of ESRI (eds. B. R. Dowling and J. Durkan)

The Irish Economy and Society in the 1980s (Papers presented at ESRI Twenty-first Anniversary Conference)
Staff Members of ESRI

The Economic and Social State of The Nation
J. F. Meenan, M. P. Fogarty, J. Kavanagh and L. Ryan

The Irish Economy: Policy and Performance 1972-1981
P. Bacon, J. Durkan and J. O'Leary

Employment and Unemployment Policy for Ireland
Staff Members of ESRI (eds., Denis Conniffe and Kieran A. Kennedy)

Public Social Expenditure – Value for Money? (Papers presented at a Conference, 20 November 1984)

Policy Research Series:

1. *Regional Policy and the Full-Employment Target* — M. Ross and B. Walsh
2. *Energy Demand in Ireland, Projections and Policy Issues* — S. Scott
3. *Some Issues in the Methodology of Attitude Research* — E. E. Davis *et al.*
4. *Land Drainage Policy in Ireland* — Richard Bruton and Frank J. Convery
5. *Recent Trends in Youth Unemployment* — J. J. Sexton

Broadsheet Series:

1. *Dental Services in Ireland* — P. R. Kaim-Caudle
2. *We Can Stop Rising Prices* — M. P. Fogarty
3. *Pharmaceutical Services in Ireland* — P. R. Kaim-Caudle
   assisted by Annette O'Toole and Kathleen O'Donoghue
4. *Ophthalmic Services in Ireland* — P. R. Kaim-Caudle
   assisted by Kathleen O'Donoghue and Annette O'Toole
5. *Irish Pensions Schemes, 1969* — P. R. Kaim-Caudle and J. G. Byrne
   assisted by Annette O'Toole
6. *The Social Science Percentage Nuisance* — R. C. Geary
7. *Poverty in Ireland: Research Priorities* — Brendan M. Walsh
8. *Irish Entrepreneurs Speak for Themselves* — M. P. Fogarty
9. *Marital Desertion in Dublin: An Exploratory Study* — Kathleen O'Higgins
10. *Equalization of Opportunity in Ireland: Statistical Aspects*
    R. C. Geary and F. S. Ó Muircheartaigh
11. *Public Social Expenditure in Ireland* — Finola Kennedy
12. *Problems in Economic Planning and Policy Formation in Ireland, 1958–1974*
    Desmond Norton
13. *Crisis in the Cattle Industry* — R. O'Connor and P. Keogh
14. *A Study of Schemes for the Relief of Unemployment in Ireland*
    R. C. Geary and M. Dempsey
    with Appendix — E. Costa
15. *Dublin Simon Community, 1971-1976: An Exploration* — Ian Hart
16. *Aspects of the Swedish Economy and their Relevance to Ireland*
    Robert O'Connor, Eoin O'Malley and Anthony Foley
17. *The Irish Housing System: A Critical Overview*
    T. J. Baker and L. M. O'Brien

18. *The Irish Itinerants: Some Demographic, Economic and Educational Aspects* — M. Dempsey and R. C. Geary
19. *A Study of Industrial Workers' Co-operatives* — Robert O'Connor and Philip Kelly
20. *Drinking in Ireland: A Review of Trends in Alcohol Consumption, Alcohol Related Problems and Policies towards Alcohol* — Brendan M. Walsh
21. *A Review of the Common Agricultural Policy and the Implications of Modified Systems for Ireland* — R. O'Connor, C. Guiomard and J. Devereux
22. *Policy Aspects of Land-Use Planning in Ireland* — Frank J. Convery and A. Allan Schmid
23. *Issues in Adoption in Ireland* — Harold J. Abramson

Geary Lecture Series:

1. *A Simple Approach to Macro-economic Dynamics* (1967) — R. G. D. Allen
2. *Computers, Statistics and Planning-Systems or Chaos?* (1968) — F. G. Foster
3. *The Dual Career Family* (1970) — Rhona and Robert Rapoport
4. *The Psychosonomics of Rising Prices* (1971) — H. A. Turner
5. *An Interdisciplinary Approach to the Measurement of Utility or Welfare* (1972) — J. Tinbergen
6. *Econometric Forecasting from Lagged Relationships* (1973) — M. G. Kendall
7. *Towards a New Objectivity* (1974) — Alvin W. Gouldner
8. *Structural Analysis in Sociology* (1975) — Robert K. Merton
9. *British Economic Growth 1951-1973: Success or Failure?* (1976) — R. C. O. Matthews
10. *Official Statisticians and Econometricians in the Present Day World* (1977) — E. Malinvaud
11. *Political and Institutional Economics* (1978) — Gunnar Myrdal
12. *The Dilemmas of a Socialist Economy: The Hungarian Experience* (1979) — János Kornai
13. *The Story of a Social Experiment and Some Reflections* (1980) — Robert M. Solow
14. *Modernisation and Religion* (1981) — P. L. Berger
15. *Poor, Relatively Speaking* (1983) — Amartya K. Sen
16. *Towards More Rational Decisions on Criminals* (1984) — Daniel Glaser

General Research Series:

1. *The Ownership of Personal Property in Ireland* — Edward Nevin
2. *Short-Term Economic Forecasting and its Application in Ireland* — Alfred Kuehn
3. *The Irish Tariff and The E.E.C.: A Factual Survey* — Edward Nevin
4. *Demand Relationships for Ireland* — C. E. V. Leser
5. *Local Government Finance in Ireland: A Preliminary Survey* — David Walker
6. *Prospects of the Irish Economy in 1962* — Alfred Kuehn
7. *The Irish Woollen and Worsted Industry, 1946-59: A Study in StatisticalMethod* — R. C. Geary
8. *The Allocation of Public Funds for Social Development* — David Walker
9. *The Irish Price Level: A Comparative Study* — Edward Nevin
10. *Inland Transport in Ireland: A Factual Study* — D. J. Reynolds
11. *Public Debt and Economic Development* — Edward Nevin
12. *Wages in Ireland, 1946-62* — Edward Nevin

13. *Road Transport: The Problems and Prospects in Ireland* D. J. Reynolds
14. *Imports and Economic Growth in Ireland, 1947-61* C. E. V. Leser
15. *The Irish Economy in 1962 and 1963* C. E. V. Leser
16. *Irish County Incomes in 1960* E. A. Attwood and R. C. Geary
17. *The Capital Stock of Irish Industry* Edward Nevin
18. *Local Government Finance and County Incomes* David Walker
19. *Industrial Relations in Ireland: The Background* David O'Mahony
20. *Social Security in Ireland and Western Europe* P. R. Kaim-Caudle
21. *The Irish Economy in 1963 and 1964* C. E. V. Leser
22. *The Cost Structure of Irish Industry 1950-60* Edward Nevin
23. *A Further Analysis of Irish Household Budget Data, 1951-52* C. E. V. Leser
24. *Economic Aspects of Industrial Relations* David O'Mahony
25. *Phychological Barriers to Economic Achievement* P. Pentony
26. *Seasonality in Irish Economic Statistics* C. E. V. Leser
27. *The Irish Economy in 1964 and 1965* C. E. V. Leser
28. *Housing in Ireland: Some Economic Aspects* P. R. Kaim-Caudle
29. *A Statistical Study of Wages, Prices and Employment in the Irish Manufacturing Sector* C. St. J. O'Herlihy
30. *Fuel and Power in Ireland: Part I. Energy Consumption in 1970* J. L. Booth
31. *Determinants of Wage Inflation in Ireland* Keith Cowling
32. *Regional Employment Patterns in the Republic of Ireland* T. J. Baker
33. *The Irish Economy in 1966* The Staff of The Economic Research Institute
34. *Fuel and Power in Ireland: Part II. Electricity and Turf* J. L. Booth
35. *Fuel and Power in Ireland: Part III. International and Temporal Aspects of Energy Consumption* J. L. Booth
36. *Institutional Aspects of Commercial and Central Banking in Ireland* John Hein
37. *Fuel and Power in Ireland: Part IV. Sources and Uses of Energy* J. L. Booth
38. *A Study of Imports* C. E. V. Leser
39. *The Irish Economy in 1967* The Staff of The Economic and Social Research Institute
40. *Some Aspects of Price Inflation in Ireland* R. C. Geary and J. L. Pratschke
41. *A Medium Term Planning Model for Ireland* David Simpson
42. *Some Irish Population Problems Reconsidered* Brendan M. Walsh
43. *The Irish Brain Drain* Richard Lynn
44. *A Method of Estimating the Stock of Capital in Northern Ireland Manufacturing Industry: Limitations and Applications* C. W. Jefferson
45. *An Input-Output Analysis of the Agricultural Sector of the Irish Economy in 1964* R. O'Connor with M. Breslin
46. *The Implications for Cattle Producers of Seasonal Price Fluctuations* R. O'Connor
47. *Transport in the Developing Economy of Ireland* John Blackwell
48. *Social Status and Inter-Generational Social Mobility in Dublin* Bertram Hutchinson
49. *Personal Incomes by County, 1965* Miceal Ross
50. *Income-Expenditure Relations in Ireland, 1965-1966* John L. Pratschke
51. *Costs and Prices in Transportable Goods Industries* W. Black, J. V. Simpson, D. G. Slattery
52. *Certain Aspects of Non-Agricultural Unemployment in Ireland* R. C. Geary and J. G. Hughes

84. *Economic Aspects of Local Authority Expenditure and Finance* J. R. Copeland and Brendan M. Walsh
85. *Population Growth and other Statistics of Middle-sized Irish Towns* D. Curtin, R. C. Geary, T. A. Grimes and B. Menton
86. *The Income Sensitivity of the Personal Income Tax Base in Ireland, 1947-1972* Brendan R. Dowling
87. *Traditional Families? From Culturally Prescribed to Negotiated Roles in Farm Families* Damian F. Hannan and Louise Katsiaouni
88. *An Irish Personality Differential: A Technique for Measuring Affective and Cognitive Dimensions of Attitudes Towards Persons* E. E. Davis and Mary O'Neill
89. *Redundancy and Re-Employment in Ireland* Brendan J. Whelan and Brendan M. Walsh
90. *A National Model for Fuel Allocation – A Prototype* E. W. Henry and S. Scott
91. *A Linear Programming Model for Irish Agriculture* Robert O'Connor, Miceal Ross and Michael Behan
92. *Irish Educational Expenditures – Past, Present and Future* A. Dale Tussing
93. *The Working and Living Conditions of Civil Service Typists* Nóirin O'Broin and Gillian Farren
94. *Irish Public Debt* Richard Bruton
95. *Output and Employment in the Irish Food Industry to 1990* A. D. O'Rourke and T. P. McStay
96. *Displacement and Development: Class, Kinship and Social Change in Irish Rural Communities* Damian F. Hannan
97. *Attitudes in the Republic of Ireland Relevant to the Northern Problem: Vol. I: Descriptive Analysis and Some Comparisons with Attitudes in Northern Ireland and Great Britain* E. E. Davis and R. Sinnott
98. *Internal Migration Flows in Ireland and their Determinants* J. G. Hughes and B. M. Walsh
99. *Irish Input-Output Structures, 1976* E. W. Henry
100. *Development of the Irish Sea Fishing Industry and its Regional Implications* R. O'Connor, J. A. Crutchfield, B. J. Whelan and K. E. Mellon
101. *Employment Conditions and Job Satisfaction: The Distribution, Perception and Evaluation of Job Rewards* Christopher T. Whelan
102. *Crime in the Republic of Ireland: Statistical Trends and Their Interpretation* David B. Rottman
103. *Measure of the Capital Stock in the Irish Manufacturing Sector, 1945-1973* R. N. Vaughan
104. *A Study of National Wage Agreements in Ireland* James F. O'Brien
105. *Socio-Economic Impact of the Construction of the ESB Power Station at Moneypoint, Co. Clare* R. O'Connor, J. A. Crutchfield and B. J. Whelan
106. *The Financing of Third-level Education* A. C. Barlow
107. *An Input-Output Analysis of New Industry in Ireland in 1976* E. W. Henry
108. *Social Insurance and Absence from Work in Ireland* Gerard Hughes
109. *The Distribution of Income in the Republic of Ireland: A Study in Social Class and Family-Cycle Inequalities* David B. Rottman, Damian F. Hannan and Niamh Hardiman, Miriam M. Wiley
110 *The Economic and Social Circumstances of the Elderly in Ireland* B. J. Whelan and R. N. Vaughan

111. *Worker Priorities, Trust in Management and Prospects for Workers' Participation*
Christopher T. Whelan
112. *The Impact of Energy Prices on the Irish Economy during 1973-1981*
E. W. Henry
113. *Schooling and Sex Roles: Sex Differences in Subject Provision and Student Choice in Irish Post-Primary Schools*
D. Hannan, R. Breen and B. Murray, D. Watson, N. Hardiman, K. O'Higgins
114. *Energy Crops, Forestry and Regional Development in Ireland*
Frank J. Convery and Kathleen Dripchak
115. *Aggregate Supply, Aggregate Demand and Income Distribution in Ireland: A Macro-sectoral Analysis*
John Bradley and Connell Fanning
116. *Social Mobility in the Republic of Ireland: A Comparative Perspective*
Christopher T. Whelan and Brendan J. Whelan
117. *Attitudes towards Poverty and Related Social Issues in Ireland*
E. E. Davis, Joel W. Grube and Mark Morgan
118. *A Study of New House Prices in Ireland in the Seventies*
Ian J. Irvine
119. *Education and the Labour Market: Work and Unemployment Among Recent Cohorts of Irish School Leavers*
Richard Breen
120. *Payroll Tax Incidence, the Direct Tax Burden and the Rate of Return on State Pension Contributions in Ireland*
Gerard Hughes
121. *Crime Victimisation in the Republic of Ireland*
Richard Breen and David B. Rottman
122. *Medium-term Analysis of Fiscal Policy in Ireland: A Macroeconometric Study of the Period 1967-1980*
John Bradley, Connell Fanning, Canice Prendergast and Mark Wynne
123. *The Irish Wealth Tax A Case Study in Economics and Politics*
Cedric Sandford and Oliver Morrissey
124. *Aspects of Freight Transport in Ireland*
Jack Short
125. *Small-Scale Manufacturing Industry in Ireland*
Kieran A. Kennedy and Tom Healy